·中英双语·

DK神秘的
树百科

[英]简·格林 著
[英]克莱尔·麦克尔法特里克 绘
安 安译

DK

黑龙江少年儿童出版社

登记号：黑版贸审字08-2020-060号

图书在版编目（CIP）数据

DK神秘的树百科：汉、英 / （英）简·格林
(Jen Green) 著 ；（英）克莱尔·麦克尔法特里克
(Claire McElfatrick) 绘 ；安安译. -- 哈尔滨：黑龙
江少年儿童出版社，2020.9（2021.4重印）
ISBN 978-7-5319-6735-4

Ⅰ. ①D… Ⅱ. ①简… ②克… ③安… Ⅲ. ①树木一
少儿读物一汉、英 Ⅳ. ①S718.4-49

中国版本图书馆CIP数据核字(2020)第113663号

DK Penguin Random House

DK 神秘的树百科
DK SHENMI DE SHUBAIKE

[英] 简·格林 著
[英] 克莱尔·麦克尔法特里克 绘
安 安 译

出 版 人　商　亮
项目策划　顾吉霞
责任编辑　商　亮　顾吉霞
出版发行　黑龙江少年儿童出版社
　　　　　（哈尔滨市南岗区宣庆小区 8 号楼　邮编：150090）
网　　址　www.lsbook.com.cn
经　　销　全国新华书店
印　　装　鹤山雅图仕印刷有限公司
开　　本　889 mm×1194 mm　1/16
印　　张　11.25
字　　数　220 千字
书　　号　ISBN 978-7-5319-6735-4
版　　次　2020 年 9 月第 1 版
印　　次　2021 年 4 月第 2 次印刷
定　　价　118.00 元

（如有缺页或倒装，本社负责退换）

Original Title: RHS The Magic & Mystery of Trees
Copyright © Dorling Kindersley Limited, 2019
A Penguin Random House Company

混合产品
源自负责任的
森林资源的纸张
FSC® C018179

FOR THE CURIOUS
www.dk.com

DK神秘的 树百科

请进入森林，一起来探索神秘的树木世界吧！你知道树木可以发送地下信息吗？你知道树木如何照顾它们的家庭吗？树木比表面看起来要复杂得多……

从最深、最茂密的森林到我们周围的城镇，树木无处不在。

我们与树木共享这个世界，与它们毗邻而居，却常常忽视它们。

翻阅这本书，你会发现树木的秘密生活。

目 录

什么是树木？

    树木是巨大的植物，高高耸立。有的树木独自生长在后花园里，也有很多树木聚集在一起，形成了茂密的森林。

    树木是自然界真正的奇观。有些种类的树木长得比五十辆汽车摞起来还要高！树木可以存活数百年。世界上最古老的树木已经存活几千年了。

    树木的各个部分各司其职，既分工又合作。从最深处的树根到最高处的树枝上最小的叶子，树木的每个部分都在努力工作，以帮助自己生存。

      当你了解这些沉默的巨人后，将会对它们有一个全新的认识……

树木的分布

从布满岩石的海岸到郁郁葱葱的山谷，几乎到处都可以看到树木。森林是很多树木一起生长的地方。

加拿大最著名的树木是**枫树**，枫树可以生产枫糖浆。

世界上最高的树木是**红杉**，它们生活在北美洲西部。

阔叶林生长在北美洲和欧洲一些气候温暖的地区。

北美洲

南美洲

北

西北　东北

西　　　东

西南　东南

南

世界上最大的雨林是南美洲的亚马孙雨林。**热带雨林**生长在靠近赤道的地方，那里一年四季都非常热。

猴谜树生长在智利，在南美洲的最南端。

图例

金合欢　白蜡树　山杨　榕树　猴面包树　桦树　巨朱蕉　雪松　复羽叶栾树　可可　椰子树　枣椰树　花旗松　榆树　桉树　无花果树　金钱松　珙桐树　华盖棕榈　蓝花楹　日本山毛榉　刺柏　木棉　贝壳杉　酸橙树　椴树

森林的种类

森林主要有三种类型：**阔叶林、针叶林和雨林**。每种森林都由不同种类的树木组成。

巨大的**针叶林**横跨北美洲北部、俄罗斯和北欧。这些地方有漫长多雪的冬季。

欧洲

亚洲

非洲

椰子的种子可以漂流数英里*，去寻找适合生长的地方。

注：英里是英美制长度单位。

贝壳杉是新西兰独有的。它们可以生存很久，长得也很高大。

桉树（也称为尤加利树）生长在澳大利亚的干燥森林中。它们的叶子常年不落。

澳大利亚

森林覆盖了地球上近三分之一的旱地。

火炬松　桃花心木　杜果树　红树　猴谜树　枫树　挪威云杉　肉豆蔻　橡树　橄榄树　松树　白杨　颤杨　面包树　红杉　红枫　花椒　橡胶树　菩提树　香肠树　纸皮桦　云杉　北美云杉　绞杀榕　柚木　伞刺金合欢

树木是如何生存的？

你从来没有见过树木吃一碗面或啃一块花生酱三明治吧？那么树木吃什么呢？只要有阳光、水和一种叫二氧化碳的气体，树木就能生存、生长，还可以给自己制作"食物"。

植物令人惊叹的"食物"制作过程被称为光合作用。

树木很坚韧，但它们需要在温暖的环境中生存。如果树叶中的水分结冰，它们就不能给自己制作"食物"了。

进餐时间

树木上的绿叶吸收阳光，然后利用阳光中的能量来混合二氧化碳和水，制作一种含糖的液体，称为树液，也就是树木的"食物"。

夏 日

阔叶树只在春季和夏季制作"食物"，因为这两个季节阳光充沛。阔叶树的树叶在秋季脱落。针叶树一年四季都长有像针一样纤细的树叶，即使在冬季，也能继续制作"食物"。

制造氧气

在忙于制作树液时，树叶会释放一种被称为氧气的气体。所有动物，包括人类在内，都需要吸入氧气，并呼出二氧化碳。如果没有像树木这样的植物，就没有可供我们呼吸的氧气。

树木的种类

世界上的树木种类繁多，将它们分门别类不是一件容易的事情。幸运的是，树木可以分为两大种类：**阔叶树**和**针叶树**。

橡树叶

阔叶树

松果

松针

针叶树

阔叶树的叶子又宽又平。阔叶树都会开花，虽然有些树木的花小得几乎看不见。它们的种子在多汁的果实里面成熟，例如李子和无花果。大多数阔叶树在秋季落叶，在春季长出新叶。

针叶树的叶子是细长形的，称为"针"。针叶树大多数属于常青树，因为它们的叶子常年不落。针叶树的种子长在坚硬的、表皮有很多疙瘩的球果里面，例如松果。

橡 树

圆球形

圆球形树木的树枝从树干均匀地向上、向侧边伸展。

枫 树

半球形

半球形树木的树枝向侧边伸展的宽度大于向上伸展的高度。

榕 树

伞 形

伞形树木的树枝向上及侧边生长，形成一个向外扩展的形状。

鹅耳枥

椭圆形

椭圆形树木的树枝向上伸展的高度大于向侧边伸展的宽度。

树干上方的多叶部分称为树冠。树冠有不同的形状。许多阔叶树的树冠又宽又圆，而针叶树的树冠常常是尖塔形的。

柳 树

垂枝形

垂枝形树木的树枝是向下垂的。

椰子树

手掌形

手掌形树木的代表是棕榈科植物，是生长在热带地区的阔叶树。与其他树木不同，棕榈科植物不长侧枝。

柏 树

圆柱形

圆柱形树木的树枝密集，向上生长。很多针叶树又高又细。

云 杉

尖塔形

尖塔形树木的树枝越往上越短。树的顶端是尖的。

树木的各个部分

树 枝

　　树干上长出大树枝，大树枝上分裂出较小的树枝，较小的树枝上分裂出细枝。树叶从细枝上长出来。在开花结果的季节，细枝上也会长出花和果实。

树 冠

　　在距离地面很高的地方，细枝和树叶相互交织，形成一层浓密的遮盖层，称为树冠。

芽

　　春天来了，叶芽和花芽裂开了，树叶和花朵从中伸展而出。

12

树干

壮实的树干从地面生长出来。它非常强壮，支撑着树枝的重量。

树根

地面下的树根把树木牢牢地固定在地里。

无论树木生长在哪里，它们都有相同的部分：树根、树干、树枝和树叶。

树桩

当树木被砍倒或者树干被折断后，留下的部分称为树桩。

树皮是覆盖在树干上的一层薄而坚韧的表皮。

隐藏的树根

在黑暗潮湿的地下世界里，树根在土壤中扩展，形成一个木质的网络。树木的三分之一都隐藏在地下。

舒适的家园

兔子、蠕虫和甲虫这样的小动物，喜欢生活在树根之间。

坚守

树根有两个主要作用：首先，它们把树木牢牢地固定在地里，这样树木就不会在暴风雨中被吹倒；其次，它们从土壤中汲取含有矿物质的水，供树叶用来制作"食物"。

延伸

有些树木有一个很大的主根称为**直根**。虽然大多数树根向侧边生长，但直根是垂直向下生长的。为了尽可能多地找到水，一棵树的树根伸展的宽度常常大于树的高度。

树木做大部分事情都很慢，但是它们喝水很快——一棵大树每天可以从土壤中汲取数百升的水。

树干

污染

树根非常敏感。它们可以察觉土壤中的污染，并且可以通过向别的方向生长，来避开污染区域。

石油泄漏

找水者

树木的主根上分裂出较小的根。在较小的根的末端有最小的根，称为**小根**。小根上覆盖着可以感知到水的细茸毛。

小根

水在树根里向上移动。

又大又强

树木的主根很强壮，是像树枝一样的木质。每个树根的尖端都有一个硬根冠，以便在生长时深入土壤。主根在土壤里可以向下延伸达1.5米。

树干和树皮

树干支撑着树枝，就像你的骨架支撑着身体一样。树干必须非常壮实，才能够支撑起所有树枝。**没有树干的树根本就不是一棵树！**

树干内部

树干的中心是心材。当树木年轻时，心材会生长。心材被边材包围着。边材中有细管道，以便将水分从树根送往树叶。

在边材和树皮之间有一层非常薄的韧皮层。韧皮层将糖分从树叶输送到树木的其他部分。

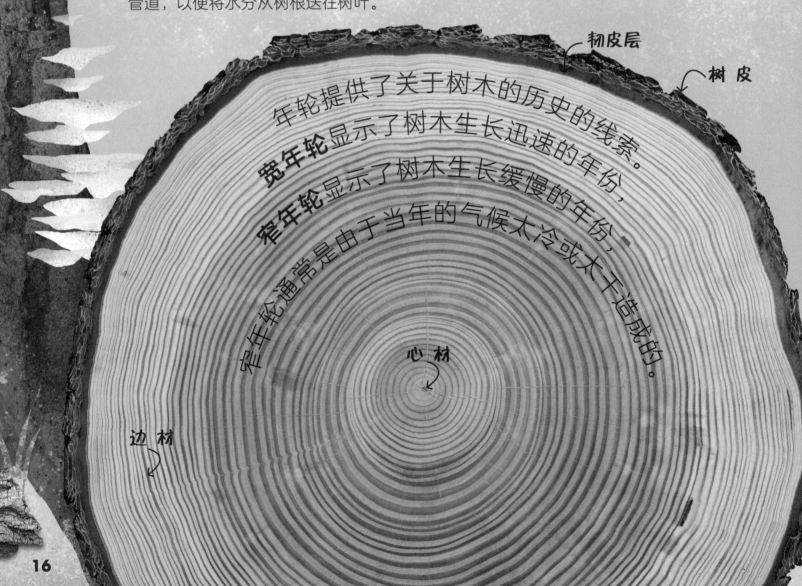

韧皮层

树皮

年轮提供了关于树木的历史的线索。宽年轮显示了树木生长迅速的年份，窄年轮显示了树木生长缓慢的年份，窄年轮通常是由于当年的气候太冷或太干造成的。

心材

边材

树皮是树干的外层，例如这棵脱皮桦树的树皮。树皮能防止树木变干，并保护树木免受昆虫和真菌的侵害。

年轻的树木有**光滑的树皮**。随着树龄变大，树皮开始裂开、脱落，有**许多皱纹**，就像这棵有鳞状树皮的树。

试着把一张纸按在树皮上，然后用蜡笔**拓印树皮的纹理**。树皮的**质地**就会在纸上显现。

不同种类的树木有不同的树皮。树皮上可能会长**地衣**，就像这些黄色的斑点。

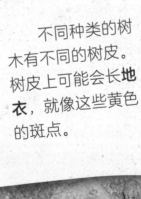

水 泵

遍布树叶的叶脉就像细小的管道，从树干边材的管道中汲取水分，并且把树叶制作的"食物"输送到树木的其他部分。

树叶

当你在户外的时候，可以仔细观察一下树叶。树叶非常特别，树木通过树叶为自己制作"食物"。

山毛榉
树叶

叶脉↰

阳光捕手

阔叶树展开它们宽阔平坦的树叶，尽可能捕捉更多的阳光。每片树叶都像一个微型太阳能电池板，吸收阳光中的能量。

树叶的形状

每种树木树叶的形状都有所不同。它们可以是又细又长的，也可以是又宽又圆的。扁平的圆形树叶擅长捕获阳光，但也会损失更多的水分。

树木不能自己从一个地方移动到另一个地方，但是它们可以非常缓慢地转动树叶，使其面向太阳。

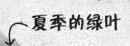
夏季的绿叶

褪色变黄

干枯的棕
色树叶

树叶是绿色的，因为它们含有一种叫作"叶绿素"的天然色素。在秋季，绿色褪去，树叶中的其他颜色呈现出来，树叶就变成了黄色、橙色或棕色。

在秋季变
成橙色

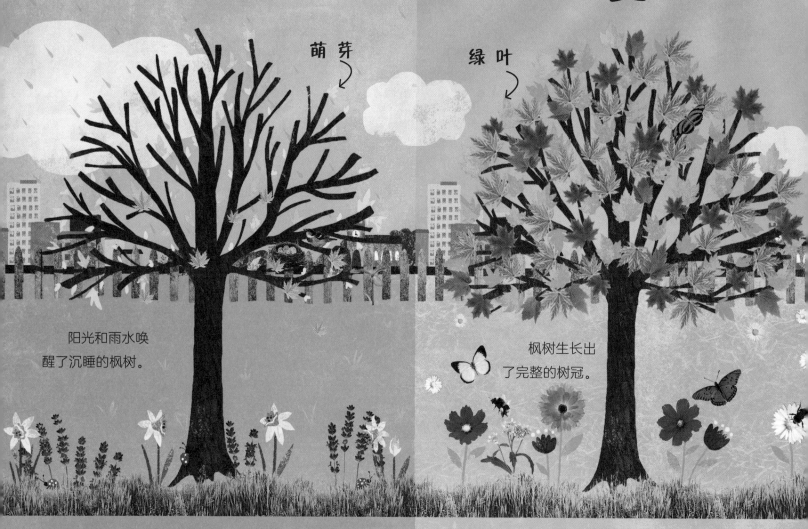

春

萌芽

阳光和雨水唤
醒了沉睡的枫树。

夏

绿叶

枫树生长出
了完整的树冠。

新芽

寒冷的冬季过后，春季是一个新的开始。天气变暖，白昼变长，树木知道冬季过去了。在春季的阳光下，树芽裂开，新叶舒展，花朵盛开。

炎热的日子

和人类一样，树木也喜爱阳光。夏季是最热的季节，有最长的白昼，树木已经准备好吸收阳光了。在这个季节，树枝上覆盖着的树叶形成了浓密的遮阳层。到了夏末，树木的果实开始生长。树木大部分是在夏季生长的。

秋

红叶

枫叶失去了绿色，纷纷脱落。

变色

到了秋季，天气转凉，白昼变短。果实成熟了，树木开始传播种子。宽阔平坦的树叶很兜风，为避免树枝被秋风吹断，树叶会脱落。在这个季节，绿叶变成黄色、橙色、红色和棕色，然后飘落到地面。

冬

没有树叶

光秃秃的树枝不容易被积雪压坏。

雪花

冬季是最冷的季节，有最短的白昼，树叶都落光了。树木看起来像死了一样，但它们只是在睡觉，等待春天。它们已经把在夏季制作的"食物"从树枝转移到树根储存起来，用以过冬。

21

花、果实和种子

　　树木需要不断繁殖新树来保持森林的健康。这就是为什么树木会开花、结果、播种。

　　花朵在春季绽放。在温暖的春日里，蜜蜂嗡嗡地叫着，从一棵树飞到另一棵树，尽可能拜访更多的花朵。

　　种子是小包裹，里面有一个蓄势待发的新生命。种子在夏季或秋季成熟。每粒种子都需要找到一个适合的地方才能生长。

　　像所有的生物一样，树木也会经历出生、生长，最终死亡的过程。但是它们给森林留下了新的希望……

左图中的可可豆荚里面含有种子。

花

想吸引**昆虫**来传播花粉的树木长有又大又艳丽的花朵，而想利用**风**来传播花粉的树木则长有极小巧的花朵。有些树木的花朵很小，很难看到，但它们同样担负着至关重要的使命。

开 花

在春季，像苹果树和樱桃树上长出了花朵的现象，称为**开花**。这些花朵使树木看起来很美丽，它们的主要作用是告诉像蜜蜂这样的昆虫：来吧！好吃的"食物"已经准备好了。

樱桃树开花

你知道花朵有雄蕊和雌蕊吗？雄蕊产生**花粉**，雌蕊产生**卵子**。花粉必须与卵子结合才能完成繁殖过程，然后发育形成种子。

小帮手

昆虫拜访花朵，是为了喝一种称为**花蜜**的甜味液体。在这个过程中，蜜蜂身上会沾到花粉，并将它带到访问的下一朵花上面，使得下一朵花受精（能够结籽）。

花粉颗粒沾在蜜蜂毛茸茸的身体上。

花朵用鲜艳的颜色和香甜的气味告诉昆虫：花蜜已经准备好了。

↰ 花粉

许多树木，像松树和冷杉树，靠风传播花粉。

果实和种子

一旦树木的花受精，**种子就开始生长了**。种子可以在果实、球果、硬壳或薄壳里生长。

针叶树种子

针叶树种子

大多数针叶树的种子长在**球果**里，而不是果实里。当种子成熟时，球果会张开，**轻盈的种子**跌落下来，并被微风吹走。紫杉和刺柏是不常见的针叶树，它们长有鸟儿爱吃的又小又苦的浆果！

樱桃

果实

如果你曾经享用过脆脆的苹果或多汁的樱桃，那就说明你已经吃过阔叶树的果实了。杧果、桃和樱桃只含有一粒大种子，称为**核**。苹果、橙子和柠檬含有许多粒小种子，称为**籽**。

这些种子的硬壳看起来大不相同，但是它们都有同样的作用：**保护种子**并帮助它们传播。

橡子

马栗子

榛子树、栗子树和核桃树的种子长有坚硬的外壳，我们称之为**坚果**。橡树的种子橡子就是一种坚果。

美国梧桐
种子

枫树和美国梧桐的种子有双"翅膀"。它们可以像小型直升机的桨叶一样旋转，飘到很远的地方着陆。

随风飘去

种子必须到达很远很宽阔的地方，以便长出新树。有些种子靠风传播。美国梧桐、枫树和白蜡树都有很轻的**有翅**种子，可以在空中旋转飞行。

随水流去

生长在河边和海边的树木会结出可以漂浮在水面上，随着水流漂走的种子。**椰子树**生长在温暖的海边。成熟的椰子掉入水中，潮水会将它们冲到远处的海岸。

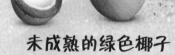

未成熟的绿色椰子

椰子很轻，可以**漂浮**在水面上。

27

动物助手

动物喜爱果实的鲜艳颜色和令它们垂涎的气味。树木把种子结在美味的果实里面，就是为了让动物帮它们**传播种子**。

探险时间

对于孩子们来说，最好的成长环境是在父母的身边，父母为我们提供需要的一切。树木则是完全不同的——它们喜欢让种子独自去遥远的地方旅行。能快速移动的动物为它们提供了完美的运输系统。

埋藏的坚果

在秋季，**松鼠**和**冠蓝鸦**通过埋藏坚果来为冬季做准备。在漫长寒冷的冬季，坚果是便利的食物储备——只要动物没有忘记把坚果埋在哪里。而被动物遗忘的坚果会在春季发芽，长成新树。

猴 粪

猴子爱吃无花果。它们的肠胃可以消化多汁的果肉，而坚硬的无花果种子会被排泄出来！

当猴子从一棵树跳到另一棵树时，它们粪便中的种子就会散布到森林中。猴粪中含有种子发芽和生长所需的全部养分。

热带无花果树

29

从种子到树木

树木是世界上最高的生物，但它们都是由很小的种子长成的。树木从种子长到最高的高度可能需要100年。下面我们来看看一棵小橡树是如何从一粒小橡子发芽，然后长得比房子还高的。

种子里面充满了养分，以保证树苗生长，直到树苗可以自己制作"食物"。

幼年的树被称为幼苗或树苗。

橡子

根

发芽
如果种子落在温暖、光照充足的潮湿土壤中，它就会开始发生巨大的变化。它会膨胀，硬壳会随之裂开，细小的根会为了汲取水分向下面延伸。

长高
一株小绿芽冒出土壤。一旦第一片树叶在阳光下展开，这棵幼小的植物就可以自己制作"食物"了。就这样，一棵新树诞生了。

橡树会在500至600年的时间里一直变粗。

长大

人类的身体在幼年时期生长，而一旦成年以后，就会停止生长。**树木则不同**——它们会继续生长。更重要的是，它们的寿命至少是人类的五倍。**100岁**的树木仍然是"青少年"！

更高更粗

随着树木变老，向上生长的速度也会逐渐变慢，当它长到**足够的高度**时就停止生长了。但是如果有空间的话，树干还会继续变粗。一棵大型老树的树干直径大约每年增长2.5厘米。

死后的"生活"

没有生物会永远存在。 树木可以存活数百年，但最终也会变老、死亡。风吹动着枯死的树，直到它的树干裂开、轰然倒地。

新家园

死亡并不是故事的结局。死去的树变成了成千上万喜欢潮湿阴暗之处的小生物的家园。诸如蛞蝓、蠕虫、潮虫、蜈蚣和蜘蛛等小动物都会搬进来住。

蜈蚣

伞菌

蟑螂

蚯蚓

毒蝇伞

生活在森林里的各种生物中，三分之一都喜欢住在腐烂的木头里或喜欢吃腐烂的木头。

养分回收

小型动物和真菌以腐烂的木头为食，并将其分解。这就使得形成树木的原材料返回土壤，为幼苗和其他植物提供了养分，给了它们一个良好的生命开端。

倒下的树木树冠上留有空隙，让阳光可以到达地面，有助于树苗茁壮生长。甚至有时候树苗会从原木中萌芽。

庭园大蜗牛

十字园蛛

蜈蚣

苔藓、蕨类植物和花朵在松软、营养丰富的腐烂原木和树桩上扎根。真菌扩展菌丝，将菌丝穿进潮湿的木头里，蘑菇便从树干上生长出来。

蛞蝓

步行虫

潮虫

甲虫在树皮下面产卵。幼虫从这些卵里孵化出来后，以腐烂的木头为食。

树木的伙伴

树木的生长和变化非常缓慢，用肉眼很难判断它们到底有多忙。年复一年，树木静默地站立不动，但实际上它们悄悄地做了很多事情。

树木是神秘的生物。最近，科学家们发现森林里的树木会互相合作：它们结交朋友，互相支持。树木会照顾它们的邻居，母树会把"食物"传送给它们的孩子和家族中的老树。

我们现在知道树木是有感觉的。它们像我们一样有嗅觉、味觉、触觉和痛感。树木可以感知危险，保护自己不受敌人伤害。我们越来越感到树木所拥有的本领远远超出我们的想象。

生活在一起

在自然界中生存很艰难。树木之间互相帮助，会更容易存活。如果一棵树遇到了麻烦，其他的树就会帮助它。树木之间通过互相合作，使森林在冬季更温暖、更有遮蔽性，在夏季更凉爽、更潮湿、更阴凉。

家庭和朋友

同一种类的树木会互相关照。橡树、山毛榉和云杉只喜欢与自己的同类分享水和"食物"。但在一些地方，不同种类的树木也会互相关照。

团结就是力量

树木长得越来越茂盛，直到碰到旁边的树木。这样就形成了一个由树枝和树叶组成的屋顶，保护森林免受风暴袭击。如果大量的树木死亡并留下空隙，强风就可能由此侵入，破坏森林。

如果一棵树受到损伤，面临死亡，那么它的邻居会传送"食物"给它，帮助它生存下去。

树根

树根在土壤中伸展，形成一张隐藏的网。森林里的邻居们彼此保持联系，并且通过它们的根来分享"食物"。

树根网络

树木之间会保持联系。 专家们发现，森林中的树木不仅通过**树根**连接，还通过**真菌**——也就是我们所说的蘑菇连接在一起。

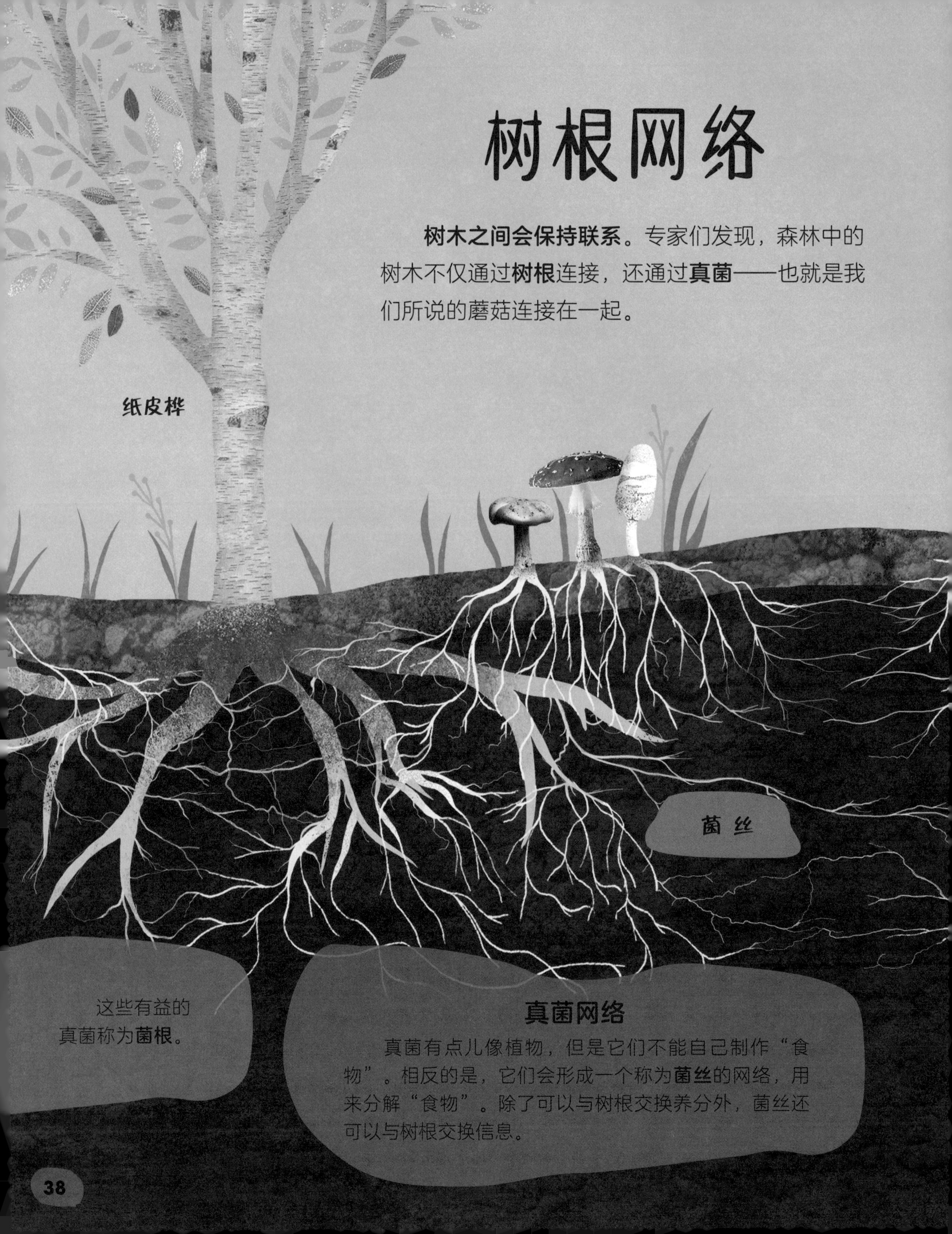

纸皮桦

菌丝

这些有益的真菌称为**菌根**。

真菌网络

真菌有点儿像植物，但是它们不能自己制作"食物"。相反的是，它们会形成一个称为**菌丝**的网络，用来分解"食物"。除了可以与树根交换养分外，菌丝还可以与树根交换信息。

有些种类的真菌能帮助树木清理污染，并且阻挡会让树木生病的其他种类的真菌。

蘑菇和伞菌是真菌的果实。

花旗松

真菌

养分

"食物"和水

树根

作为对真菌网络的回赠，树木给真菌提供水和"食物"。

森林大家庭

同一种类的树木就像一个家庭。当树苗陷入困境时，它的母亲会来帮助它。那些很老或受损的树木也不会被遗忘。即使是最强壮的树木也有可能受到疾病或昆虫的侵袭，不时地需要帮助。通过互相帮助，使得整个森林保持健康。

小树苗

母亲的关爱

成熟的树木总是照顾着较年轻和较年老的树木。许多幼树直接在母树下发芽。因为幼树在母树的阴影下不能很好地生长，所以母树就把自己的树液和养分传送给幼树，让它们保持活力，直到它们长得足够高，可以吸收到阳光。

年长的树木经常会挡住阳光，使其不能直接照射幼树。这不仅没有害处，反而有益处，因为在年轻的时候缓慢地生长，会使树木更长寿。

援助之手

这个树桩没有树叶，无法制作"食物"，但是它仍然以某种方式存活着，因为森林中的其他树木正在通过树根网络喂养这个树桩。它甚至有可能是某棵成年树的母亲。

没有树叶的树桩可以存活数百年。

树木的感觉

树木没有眼睛、耳朵、手指和脚趾。它们与我们如此不同，所以在很长一段时间里，没有人知道树木是有感觉的。但是现在我们知道了，树木可以辨别周围的事物。

它们可以感知外面的世界……

热和冷

树木可以感知热和冷。即使是很小的种子也知道气温是否已经变得足够温暖，可以开始发芽、生长。

味觉

当动物啃食树叶时，树木也可以尝到动物的唾液！树木甚至可以分辨不同动物的唾液味道。

触觉

树根非常敏感。树木可以辨别在地下缠结的树根中哪些是自己的。它们还可以辨别邻居和自己是否是同一种类的树木。

听觉

树根可以听到流水的声音并向着流水的方向生长。即使水源被完全封闭，树木也知道水在哪里。我们还不清楚树木是如何做到这一点的。

我们才开始揭开关于树木感知的神秘面纱。

视觉

树木没有眼睛，却能感知光并向着光线的方向生长。树木上的每片树叶都可以分辨光线来自哪个方向。毕竟，它们需要阳光来制作"食物"。

时间感

虽然树木的生活节奏缓慢，但它们还是会记住时间的。在春季，树木感知白天变长了。到了秋季，它们知道白天变短，需要准备过冬了。

......甚至可以回应。

做自己

没有两棵树是完全一样的。即使处于相同的条件下，树木也会各自长成不同的形状。有些树木参与树根网络，有些树木则是"独行侠"。

交流

树木可以与昆虫交流。花朵的颜色和香味就像是广告，告诉蜜蜂和蝴蝶：这里有"食物"。

"喊叫"

当树木口渴时，它们会开始"喊叫"！如果水不能从根部流到树叶上，树干就开始抖动——这是树木的抱怨方式。

来自中国的**珙桐树**（手帕树）上的花朵看起来就像晃来晃去的餐巾纸。

科学家们开始研究树根是如何像大脑一样"思考"的。

树木的防御

想象一下，如果你是一棵树，一只昆虫开始咬你——哎哟！好痛！幸运的是，树木有很多聪明的方法来阻止昆虫入侵。

黏稠的树液

攻击**美国黑松**的甲虫发现自己陷入了黏稠的困境！在甲虫享用美味的树皮之前，黑松就分泌出黏稠的**树液陷阱**，将甲虫束缚起来。

松甲虫

苦涩的晚餐

山毛榉、橡树和云杉可以把一种称为"**单宁**"的化学物质注入它们的叶子中，使其味道变得苦涩。这破坏了昆虫的美食，所以它们就得转移到另一棵树上。

树木可以分辨它们被哪种虫子咬了，并且有针对性地寻找帮手。

橡树叶
卷蛾毛虫

昆虫救援者

树木用不同的策略来保护自己免受喜爱吸食树液的**蚜虫**的侵害。这些树木释放出召唤**瓢虫**的特殊气味，那些长有斑点的瓢虫就会飞过来追捕蚜虫。如果赶巧，它们每天都可以吃到蚜虫。

正在吃蚜
虫的瓢虫

遭到攻击

长颈鹿是金合欢的天敌。幸运的是，金合欢准备了一些聪明的防御措施。它们不仅能保护自己，而且还能**向其他树木示警。**

金合欢
的刺

危 险

金合欢可以感知长颈鹿的唾液。当金合欢的叶子被长颈鹿啃食时，它们就开始制造使叶子味道变得苦涩的化学物质——呸！好难吃！

小心长颈鹿

金合欢用又长又尖的刺来抵挡动物，但是长颈鹿柔韧的长舌头可以躲开尖刺，摘下一簇簇绿叶。

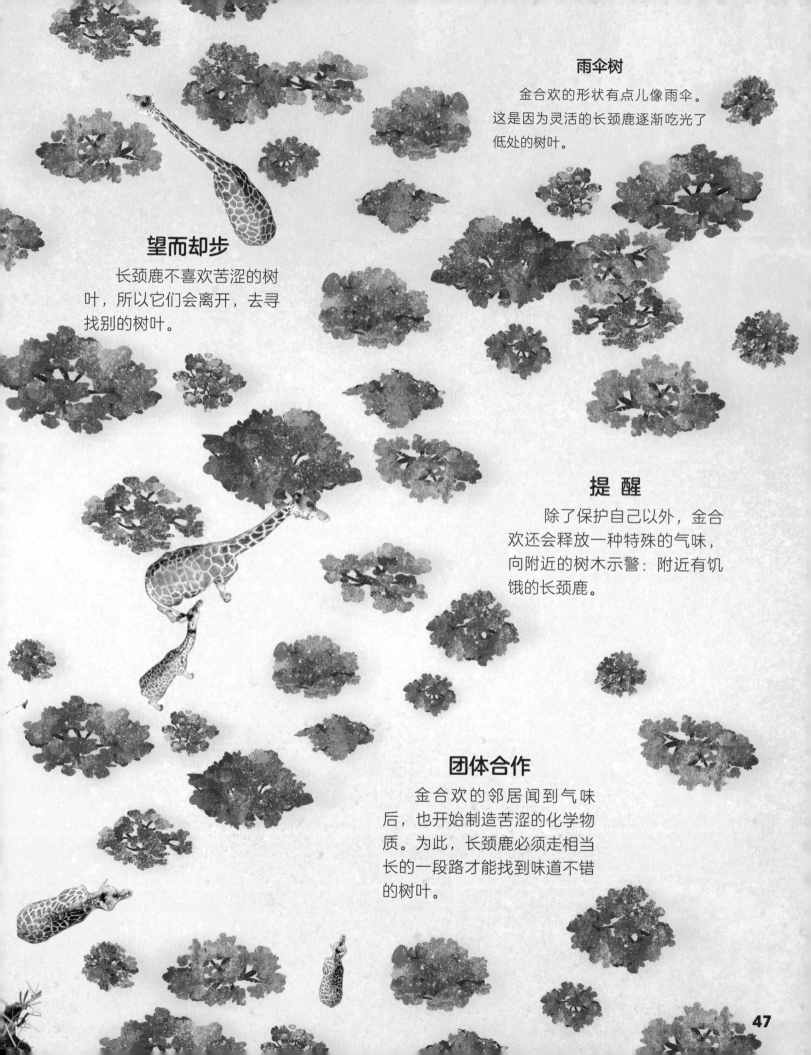

雨伞树

金合欢的形状有点儿像雨伞。这是因为灵活的长颈鹿逐渐吃光了低处的树叶。

望而却步

长颈鹿不喜欢苦涩的树叶，所以它们会离开，去寻找别的树叶。

提 醒

除了保护自己以外，金合欢还会释放一种特殊的气味，向附近的树木示警：附近有饥饿的长颈鹿。

团体合作

金合欢的邻居闻到气味后，也开始制造苦涩的化学物质。为此，长颈鹿必须走相当长的一段路才能找到味道不错的树叶。

不可思议的树木

树木是地球上最大、最重、最古老的生物。它们可以生活在安第斯山脉多雪的高地上，甚至可以克隆自己。

最大的树

最粗的树

最古老的树

世界上最高的树是生长在美国加利福尼亚州的一棵**红杉**，名为"亥伯龙神"。这棵树高达115米，大约是20只长颈鹿摞在一起的高度！

名为"谢尔曼将军"的**巨杉**是世界上最大的个体生物。它巨大的树干直径超过8米。

墨西哥的一个教堂墓地里耸立着一棵蒙特祖玛落羽**杉**，名为"图勒树"。它是世界上最粗的树：树干的周长有36米，树高却只有35米。

世界上已知最古老的树是一棵名为"玛土撒拉"的**狐尾松**。它生长在美国加利福尼亚州高高的怀特山脉中，已有4800年的历史。

树木可以在各种各样的地方生存很长时间，因为它们善于适应不同的环境。

最高处的树

最重的树

最古老的树根

生长在澳大利亚塔斯马尼亚岛的一棵**水松**的树根有着史诗般的10500年的历史。

在美国犹他州的森林中，有一片**颤杨**，它们都是从同一棵树上发芽生长出来的，所以它们的树根是相连的。这使得这片树成为世界上最重的生物。这片树的总重量大约是33头蓝鲸的重量。

南美洲坚韧的**多鳞树**可以在高达5200米的安第斯山脉中生长。那里非常寒冷，而且风很大，其他树木无法生长。

树木的栖息地

　　树木对于它们生存的地方并不那么挑剔。它们所需要的只是阳光和水，还有可以让它们扎根的泥土。这就是森林能遍布世界各地，并在森林深处有丰富物种的原因。

　　树木是坚韧的幸存者。分散的树木可以在陡峭多石的山上扎根，可以在遭受风暴袭击的海岸边生长，也可以在雪原的狂风中生存。树木可以在黄沙飞扬的沙漠边缘存活，也可以生长在城市中心繁忙的街道边。事实上，树木几乎可以在任何地方生存。

　　本章将介绍树木在世界各地的栖息地。请继续阅读，你将学到树木是如何为哺乳动物、鸟类、昆虫和鱼类提供栖息地的。

树木家园

这棵大橡树是许多生物的家园。从最上面的树枝到最深处的树根，许多动物和植物都生活在一起，就像生活在同一座公寓楼里的人一样。

猫头鹰和蝙蝠等肉食性动物正在寻找猎物。

尽情地吃吧！

像这样的老树可以为大家提供食物，就像一家天然的**超市**。

昆虫、鸟和鹿这样的植食性动物，正用力咀嚼着树木的各个部分。它们啃食树木的叶子、嫩芽、果实，甚至树皮。

在地下，树木的地下室中，**蠕虫**、**甲虫**和**真菌**在树根间寻找食物。

乌鸫在高处的树枝上建造舒适的巢来产卵。巢可以为雏鸟保温并保护它们免受狐狸等**天敌**的侵害。

这只**啄木鸟**用它超级锋利的尖喙敲打树皮，挖出多汁的幼虫，然后用黏糊糊的舌头把它粘出来吧嗒吧嗒地吃掉。

秋季，**松鼠**把坚果埋在地下，为过冬储备食物。

狐狸用它们锋利的爪子在树根间挖出舒适的**洞穴**。

热带雨林

潮湿的亚马孙热带雨林是世界上最大的生物家园，其中的一小片就有数百种不同种类的树木。

露生层

被称为"露生层"的非常高的树木，冒出雨林的树冠不必与其他树木竞争就可以享受阳光。

树冠

雨林里的树木喜爱炎热潮湿的天气。它们长得很高，形成了一个密实的遮阴屋顶。

蓝闪蝶

木棉树在雨林中很常见。它们可以长到70米高。

攀缘的藤蔓

醒目的喙

巨嘴鸟长有颜色鲜艳的喙，喙的长度可达19厘米。

红绿金刚鹦鹉

红嘴镰嘴犁雀

长长的四肢
蜘蛛猴用它们长长的胳膊、腿和尾巴在树林间荡来荡去。

蟒蛇

蜈蚣

瓶状叶

金镖蛙

小丑蟾蜍

狼蛛

诡秘的大猫
美洲虎身上的斑点使它可以藏身于树叶斑驳的阴影中。当美洲虎行猎时，会悄悄地在树林中潜行。

二趾树懒

不见外的客人
附生植物是生长在其他植物上的植物。雨林中的树木抚育了很多附生植物。

附生植物

翡翠树蚺

凤梨

倭儒狨猴

棕榈

麝雉

下层
下层是昏暗的，因为上面的树冠遮挡了大部分的阳光。

森林的地面
雨林的地面是黑暗而且干燥的，因为高层几乎吸收了所有的阳光和水。

巨型食蚁兽

55

温带雨林

与热带雨林不同，**温带雨林生长在气候温和的区域**，气温既不太热，也不太冷。这里的树木大多数是针叶树。这种雾蒙蒙的森林是很多喜欢凉爽环境的动物的家园。

河 狸

河狸在雨林的溪流中修筑水坝。它们用锋利的门牙啃**树苗**。倒下的树成了水坝的一部分。

火 草

巨型红杉

大 猫

美洲狮是一种大型猫科动物。它们通常会悄悄地在雨林中潜行，并且暴起突袭像鹿那样的大型动物，咬住猎物的颈部以杀死它们。

黄花水芭蕉

树 蛙

太平洋树蛙生活在雨林的池塘和潮湿的沟渠中。它们的颜色呈棕色、灰色或绿色，并可以通过**改变身体的颜色**，与周围的环境融为一体。

温带雨林印象

世界上最大的温带雨林生长在北美洲的西海岸。它主要由**针叶树**组成，包括世界上最高的树——巨型红杉。那里的树木生长得很好，因为它们有充足的水分。冬季雨量大，夏季有从海洋滚滚而来的**浓浓的湿雾**。

温和的巨人

驼鹿是最大的鹿。雄性驼鹿长有**巨大的多叉鹿角**。它们在树木之间穿梭，贪婪地咀嚼蕨类植物、草、嫩枝和树皮。

俄勒冈葡萄

厚脸皮的花栗鼠

花栗鼠是一种敏捷的小型啮齿动物。它们在森林里跑来跑去，寻找坚果、水果和种子，把食物放在它们隆起的**颊囊**里带回洞穴。

浣 熊

这种好斗的哺乳动物的眼睛上有一条宽宽的黑色条纹，就像**强盗的面具**。它们的行为也像强盗一样，经常袭击鸟巢、偷蛋。

杜鹃花

沼泽森林

在炎热潮湿的沼泽里，不寻常的红树生长
在流动缓慢、混浊的水中。在这个沿海热带森
林里，树木每天会被咸潮冲洗两次。

长鼻猴

孙德里红树林

踩高跷的树木

红树长有很高的树根，能把树干抬
出水面。这些树根的作用是过滤掉水中
大部分的盐，让树木喝到淡水。

雄招潮蟹

掉落的树叶被螃蟹抢着吃掉。

咸水鳄潜伏在暗处，等待猎物。
红树根部周围的泥浆是这种世界上最
大的爬行动物的完美藏身之地。

马来渔鸮

有些红树把盐分转移到树叶上，树叶就会脱落。

孟加拉国的孙德尔本斯红树林是世界上最大的红树林。

蓝翡翠

孟加拉虎在沼泽地里潜行，寻找鹿。它们身上的条纹为它们在水边高高的芦苇丛中行动提供了完美的伪装。

弹涂鱼是一种奇怪的鱼类，它们可以离开水生存，并且可以用强壮的前鳍在泥地上滑行。

很多鱼藏身于树根之间。

红树的根扎进泥浆里，减缓水的流动，防止海岸随着时间的推移被侵蚀（被海潮冲走）。

幽灵般的**乌林鸮**在森林上空静静地飞行，观察地面上的猎物踪迹。

云杉

落叶松

交嘴雀长有钳形喙，尖端有重叠，可以撬开松果，吃里面的松子。

熊用锋利的爪子爬树，打开蜂巢，咕咚咕咚地喝蜂蜜。

红交嘴雀　　熊

雄性**驼鹿**长有巨大的多叉鹿角，喜欢在树上蹭鹿角。

雪 林

浓密黑暗的**泰加森林**在冬天被雪覆盖。它跨越加拿大、俄罗斯和北欧，像一条巨大的绿色围巾一样围绕着世界。

凉爽的针叶树

泰加森林主要由云杉和冷杉这样的**针叶树**组成。针叶树长有细尖形、滑得像蜡一样的树叶，有助于树木摆脱积雪。

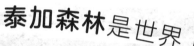

泰加森林是世界上覆盖面积最大的森林。

驯鹿在有遮蔽的森林中度过冬天，以树枝上和积雪下面的地衣为食。

纸皮桦

驯 鹿

松貂在白雪皑皑的森林地面上飞奔，利用嗅觉探测可以猎捕的小动物。

山野兔

白色世界

北方的森林夏季短暂而凉爽，冬季漫长而严酷。积雪经常连续数月覆盖地面。

极端幸存者

　　大多数树木喜欢生活在温暖、阳光明媚的地方。然而，一些有耐力的树木可以应付难以置信的**恶劣条件**，比如极端**炎热**、极端**寒冷**或长期干旱的环境。

长在边缘

松树和柏树可以生长在多风的悬崖上。强大的风吹断了树枝，只留下树的一个侧边。我们通过这些树木的形状，就可以辨别风向。

低矮的柳树

泰加森林的北部是冰冻的北极苔原。在那里，灌木状的北极柳可以适应严寒和积雪。它们匍匐在地面上，躲避呼啸的寒风，需要100年才能长到20厘米高。

树木喜欢水，不过有些树木已经适应了非常
干旱的环境，它们竭尽所能地汲取每一滴水。

树袋熊不能从桉树叶
中获得充足的养分，所以
它们每天睡觉的时间长达
18小时，以便节省能量。

树袋熊的家园

高大的桉树生长在澳
大利亚的干旱地区。树袋
熊只吃它的树叶。

水 罐

猴面包树生长在非洲
干燥的稀树草原上。在雨
季，猴面包树把水储存在
树干里，用来度过旱季。

耐 火

西黄松可以在火灾中幸
存。这是因为它们有厚达10
厘米的保护性树皮。

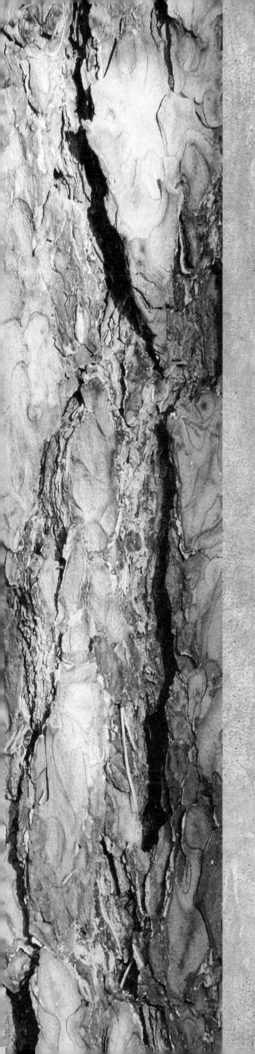

树木和我

　　树木以各种各样的方式帮助人类。它们给我们提供食物、木材和许多其他方面的产品。它们制造氧气，使空气变得新鲜。它们帮助我们的星球保持干净的生活环境。

　　树木很好地照顾着我们和其他生物。然而，在世界各地，人们正在伤害树木。人们为了获取木材，为了给农场和城市腾出空间，大肆砍伐森林。人们还制造污染，危及树木的生命。

　　树木做了很多事情来保护世界的绿色和健康。它们为无数植物、动物以及数百万生活在森林中的人类提供了家园。作为回报，我们每个人都应该好好保护树木。

冰淇淋

垃圾食品

棕榈油可用于制作**比萨饼**、**饼干**和**冰淇淋**。热带雨林正在遭到砍伐，目的是腾出空间种植产油的树木。为了保护热带雨林，最好不要吃太多用棕榈油制成的食物。

比萨饼

曲奇饼

杧果

牛油果

桃

梨

水果沙拉

以前的果树是野生的，但是现在的果树主要由人类种植在**果园**或**种植园**里。我们食用的有些是水果里面的种子，例如石榴。

橙子

杏

枣

石榴

无花果

枣椰树上结的甜甜的、黏手的枣。

樱桃

收获时节

从多汁的水果到辛辣的咖喱，树木提供了很多令人垂涎的食物。没有树木，就没有巧克力、杧果酸奶和牛油果吐司！树木甚至能产出治病的药物。

苹果

杏仁

脆脆的坚果

许多坚果产自树木，包括巴西栗，它产自亚马孙热带雨林中最高的树木。椰子树生产的坚果是所有树中最大的。

阿司匹林

树木医生

治头痛的药物阿司匹林最初取自柳树皮。金鸡纳树也很有用，它可以生产一种名为"奎宁"的药物，这种药物有助于治疗疟疾。

巴西栗

腰果

李子

椰子

香味

如果没有树木，我们的食物会变得平淡无味。肉桂和肉豆蔻这样的香料真的可以让我们的饭菜变得更好吃！可乐果可以给可乐汽水调味。而味道独特的巧克力则产自可可树上的可可豆。

肉桂面包

撒在奶油蛋挞上的肉豆蔻香料

可乐

枫糖浆

口香糖

糖胶树的树液可以用来制造胶质的口香糖。

巧克力

甜蜜的糖浆

煎饼上的枫糖浆是用糖枫的树液制成的。把一个金属水龙头敲入糖枫的树干，就可以收集滴出来的树液。

纸制品

纸和卡片是用软性木屑制成的。先把碎木屑与水混合，制成糊状纸浆，然后把它碾平，干燥后就是纸。

木制品世界

看看你的家里能找到多少木制的物品？从用雪松制成的铅笔到松木家具，家里到处都有木制品。

松树成材需要至少30年的时间。

树木制品

木材是一种"超级材料"。它坚固、美观，可以用环保的方式种植。我们可以使用这种材料来建造我们的房屋，制作各种物品。

栓皮栎的树皮可以用来制作记事板、酒瓶塞和餐垫。栓皮栎的树皮很轻，可以浮在水面上。

耐磨的硬木材

像橡树和枫树这样的树木可以制造坚固的硬木，用来制作屋顶横梁和家具。桃花心木和蔷薇木这样的热带硬木可以制作吉他，因为它们可以发出清晰的声音。硬木很坚韧，但这样的树木生长得非常缓慢。

橡 胶

橡胶树渗出的乳白色汁液可以用来制成汽车或自行车的轮胎。橡胶靴、松紧带和派对上的气球都是用橡胶制成的。

一种叫作沉香的香料取自沉香树，比黄金还贵！

69

树木和地球

树木对地球上的每个人都至关重要。树木让地球保持凉爽和湿润，使它成为一个宜居的好地方。树木爱汲取水分，不过不要担心，树木愿意与我们分享。它们还帮忙制造了云彩。这个过程被称为"水循环"。

树叶接住雨水，并且把不需要的水分以水蒸气的形式释放掉。

水循环

水在海洋、空气和陆地之间循环就是"水循环"。树木在其中起着非常重要的作用。树木把水分释放回空气中，并通过树根吸水，阻止雨水直接流进大海。

海洋中的水蒸发，变成一种叫作水蒸气的气体。

森林有助于冷却地球上的空气。地球上的气候已经偏暖，如果没有树木，地球将会变得更热。

在高空中，来自树木的水蒸气形成云。

当云飘过高山时，水就会以雨或雪的形式落下。

雨

云彩向内陆移动，把雨水带到需要的地方。如果没有树木，沙漠将会覆盖地球上更多的区域。

水流回到较低的地方。

没有被树木吸收的雨水会渗入土壤或流入河流。

河流

处于危险之中的树木

在世界各地，为了利用森林下的土地，人们正在砍伐森林。 即使像亚马孙雨林这样辽阔的森林也正在缓慢缩小。人们还制造污染来伤害树木。我们需要给珍贵的森林更多地保护。

森林的毁灭

为什么森林正在消失呢？

因为人们需要获取木材，用于取暖和烹调。人们砍伐森林，目的是腾出土地来建造新的道路和城市，建造新的农场来种植庄稼，建造新的牧场来养牛。

失去的家园

当伐木工进入森林后，空气中充满了电锯的呜呜声，巨大的树木哗啦啦地倒下，原木被装载到卡车上，剩下的只有残破的树桩。无家可归的动物纷纷逃走，它们的栖息地就这样被摧毁了。

影 响

  是树根将土壤联系在一起的。如果
没有树木降低雨速和吸收雨水，那么雨
水就会将森林里的土壤冲到河里，可能
会导致洪水泛滥，最终整个地区的气候
会变得更干燥，无法种植农作物。

污 染

  城市里的汽车、工厂和发电厂可能会对
树木造成伤害。这些汽车、工厂和发电厂排
放的烟雾随风飘去，污染远处的森林，使树
木落叶、生病，最终死亡。

帮助树木

树木需要我们的关爱。它们为保护地球的健康做出了很大的贡献。作为回报，我们所有人都应该更好地保护世界上的树木。我们可以使用再生纸，还可以种一棵树。

种 树

买一棵树苗，在你家附近的花园、公园或其他开阔的空地，为树苗建造一个家园。

铁锹

树苗

挖 土

在地上挖一个深坑，确保坑的宽度是所有树根直径的两倍，深度与树根的长度相等。不同的树木喜欢不同的土壤，但大多数树木喜欢松软、潮湿的土壤，这样它们的根更容易生长。

移 植

把从花店里买来的树苗放在坑的中心，然后在坑上面放一块木板，跨过树苗的根部土壤和坑的两个边缘，这样我们就能知道坑的深度是否合适。树根的上沿应该与地面一样高。

在一些地方，比如国家公园和保护区，树木得到了保护。护林员们负责照看这里的树木，并且在需要的时候种植新树。

木桩

填土

把土填入坑中，不要压得太实。让土完全覆盖树苗的根部，并要让它有生长的空间。把树苗绑在一根木桩上，以帮助树苗在风中站立。

浇水

给你的树浇水，让它好好生长，并清除周围的杂草，驱除害虫。两到三年后，你就可以移掉木桩，因为那时你的树已经可以自己骄傲地站立了。

一些主要词汇

阔叶树

有又宽又平树叶的树木。阔叶树的果实内部含有种子。

二氧化碳

空气中的一种气体。植物用二氧化碳来制作"食物"。

叶绿素

树叶中的绿色色素。叶绿素可以吸收阳光。

克隆

动物或植物的相同副本。

针叶树

种子长在球果里的树木种类。

树冠

树木的多叶部分。

干旱

长时间没有降雨。

赤道

环绕地球中心表面的假想的圆周线。

受精

令动物或植物能够产生后代的时刻。树木的花受精后，才会结出种子。

真菌

一种生物，包括蘑菇、毒菌和霉菌。

幼虫

看起来像虫子的小昆虫。

栖息地

植物或动物的天然家园，例如森林或草地。

心材

树干的中心部分，形成于树木的生长初期。

菌丝

可以让真菌进食的微小的分枝丝。

红树

生长在沿海沼泽地区的树木。

矿物质

植物从土壤中获取的天然物质。

花蜜

花朵产生的含糖的液体，用来吸引昆虫。

养分

营养物质。植物从土壤中获取养分。

氧气

空气中的一种气体。所有生物都需要氧气才能生存。

韧皮部

树皮下的一层，负责运送养分。

光合作用

树叶把阳光和二氧化碳气体转化为有机物质和氧气的过程。

花粉

花粉是一种细小的颗粒，与植物的卵子结合后可以结出种子。

捕食动物

猎捕其他动物作为食物的动物。也被称为食肉动物。

猎物

被其他动物捕食的动物。

繁殖

生物繁衍后代的过程。

细根

细小的覆盖着细茸毛的树根。

边材

树干中靠近树皮的部分，是在最近几年生长的。

种子

能够萌发成新的植物体的部分。

幼苗

幼小的植物或树。

直根

树木垂直向下扎进土壤的主根。

叶脉

将水输送到叶子周围并且帮助叶子保持形状的管状构造。

中英词汇对照表

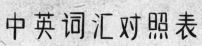

英 文	中 文	页 数
a little	少许	16
a third of	三分之一	7
about	大约, 周围	31
above	在……之上	5
absorb	吸收	55
Acacia	金合欢	6
acorn(s)	橡子	26
across	越过, 横切	7
act like	行为举止像	40
adapt(ing)	适应	49
adult(s)	成人	31
adventure	冒险	28
Africa	非洲	7
After	在……后	20
agaric	木耳	32
agarwood	沉香木	69
air	空气	9
alive	活着	37
allow(s)	允许	33
almost	几乎	6
alone	独自	5
along	沿着	57
already	已经	71
amazing	令人惊叹的	8
amazingly	令人惊讶地	42
Amazon	亚马孙	6
among	在……之间	14
Anaconda	蟒蛇	55
Andes	安第斯山脉	48
animal(s)	动物	9

英 文	中 文	页 数
antler(s)	多叉鹿角	57
anywhere	任何地方	51
apart	分别	42
aphid(s)	蚜虫	45
apple	苹果	24
Apricot(s)	杏	66
aquilaria tree(s)	沉香树	69
Arctic	北极	62
area(s)	地区	56
arm(s)	手臂	55
around	周围, 各处, 围着, 环绕	2
array	大群	51
as long as	只要	8
Ash	白蜡树	6
Asia	亚洲	7
Aspen	山杨	6
aspirin	阿司匹林	67
assistant(s)	助手	28
at least	至少	31
attack(ed)	攻击	40
attract	吸引	24
Australia	澳大利亚	7
autumn	秋, 秋天, 秋季	9
Avocado(s)	牛油果	66
avoid	避开	15
back	后面的, 返回	5
bad	坏的	47
bald cypress	落羽杉	48
balloon(s)	气球	69
bandit	强盗	57

英 文	中 文	页 数	英 文	中 文	页 数
Bangladesh	孟加拉国	59	Birch	桦树	6
Banyan	榕树	6	Bird(s)	鸟	26
Baobab	猴面包树	6	biscuit(s)	饼干	66
Bare	裸的	21	bit	一点儿	38
Bark	树皮	13	bit(ing)	咬	45
basement	地下室	52	bitter	苦的	26
bat(s)	蝙蝠	52	black	黑色，黑色的	57
batter(ed)	接连猛打	51	Blackbird(s)	乌鸫	53
beak	喙	53	Black-capped kingfisher	蓝翡翠	59
Bean(s)	豆	67	blade(s)	桨叶	27
Bear(s)	熊	60	bland	平淡无味的	67
beast(s)	动物	32	blanket	遮盖层	12
beautiful	美丽的，美观的	24	bleak	寒冷的	28
beaver(s)	河狸	56	blend	混合	56
become(s)	变成	17	block	挡住	41
bee(s)	蜜蜂	23	bloom	盛开	20
Beech	山毛榉	18	blossom	开花	24
beetle(s)	甲虫	14	blotchy	有斑点的	17
before	在……以前	7	blow	吹	26
beginning(s)	开始	20	Blow(n)	吹	27
behind	在……之后	13	Blue morpho butterfly	蓝闪蝶	54
below	在……下面,下面	14	blue whale(s)	蓝鲸	49
bendy	柔韧的	46	blustery	大风的	21
beneath	下面	72	body	身体	16
Bengal tiger	孟加拉虎	59	book	书	2
berry(berries)	浆果	26	boot(s)	靴子	69
best	最好的	28	born	诞生	30
Between	在……中间	16	bottle	瓶子	69
Beware	小心,谨防	46	bowl	碗	8
Beyond	在远处	62	brain	大脑	43
bicycle	自行车	69	branch	树枝	5
big	大	14			
big(gest)	最大的	6			

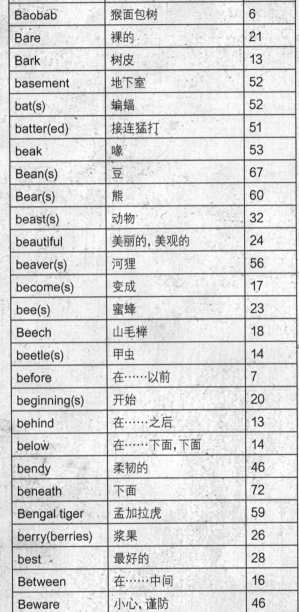

英　文	中　文	页　数	英　文	中　文	页　数
brand new	全新的	74	call	把……称为，召唤	24
brazil nut(s)	巴西栗	67	called	被称为	8
break(s)	折断，打破	13	camouflage	伪装	59
break down	分解	33	Canada	加拿大	6
breathe	呼吸	9	canopy	（树）冠，冠层	12
breeze	微风	26	cap	（根）冠	15
bright	鲜艳的	25	capture	捕获	18
bring (ing)	带到	71	car(s)	汽车	5
bristlecone pine	狐尾松	48	carbon dioxide	二氧化碳	8
broad	宽的，宽阔的	11	card	卡片	68
Broadleaved forest(s)	阔叶林	6	care(caring)	关心，关爱	40
			Carnivore(s)	肉食动物	52
Broadleaved tree(s)	阔叶树	9	Carried	被载运的	27
			carry on	继续	31
broken	残破的	72	carry	携带	16
Bromeliad	凤梨	55	case	壳	30
brown	棕色，棕色的	19	Cashew(s)	腰果	67
bud(s)	萌芽，芽，幼芽	12	cat	猫	55
Buffy fish owl	马来渔鸮	59	catch(ing)	捕获，接住	67
bug	虫子	45	catcher(s)	捕手	18
build	建造，修筑	53	cattle	牛	72
bulging	隆起的	57	cause	引起	65
bun	小圆面包	67	Cedar	雪松	6
Buried	埋藏的	28	Centipede	蜈蚣	32
burrow	打洞，洞穴	5	centre	中心	16
burst	裂	12	certain	某些	12
busy	忙碌的	9	certainly	当然	43
butterfly (butterflies)	蝴蝶	43	chainsaw(s)	电锯	72
			chance	机会	45
buzz	嗡嗡地叫	23	change(s)	变化，改变	30
Cabbage tree	巨朱蕉	6	Change (changing)	改变	21
California	加利福尼亚州	48	Cheeky	厚脸皮的	57

英 文	中 文	页 数	英 文	中 文	页 数
create	形成, 产生	11	depth	深度	74
creature(s)	生物	14	depth(s)	深处	51
Creeping	缓慢行进的	54	desert(s)	沙漠	51
crocodile	鳄鱼	58	deserve	值得	74
crop(s)	农作物	72	destroy(ed)	破坏	66
crown	冠	11	destruction	毁灭	72
crunchy	脆脆的	26	Deter(red)	却步, 被阻止	47
current	水流	27	develop	发育	26
curry	咖喱	66	devour(ed)	吞食	44
custard tart	奶油蛋挞	67	die	死亡	23
cut	砍	13	different	不同的	7
cycle	循环	70	dig	挖	53
Cypress	柏树	11	digest	消化	29
dam(s)	水坝	56	dim	昏暗的	55
damage	损坏	21	dinner	晚餐	45
damp	潮湿的	14	direction	方向	15
danger	危险	35	directly	直接地	40
dangling	悬挂的	43	disappear (ing)	消失	72
dark	阴暗的, 黑暗的	12	discover	发现	2
dart	镖	55	disease	疾病	40
Date palm	枣椰树	6	distant	遥远的, 远处的	27
day	天, 白天, 日子	9	divide	分裂	12
dead	死的	21	doctor	医生	67
death	死亡	32	dodge(s)	躲闪	46
deep	深的	62	double	双	27
deep(est)	最深的	2	Douglas fir	花旗松	6
deer	鹿	52	down	下	13
defend	防御	35	downwards	向下	11
delicate	精巧的	24	draw	吸引	14
delicious	美味的	28	dry(dried)	使变干	68
den(s)	窝	53	dry(drier)	更干燥的	73
dense	密集的	12	drift	漂移	21
dense (densest)	最茂密的	2	drink	喝	15

英 文	中 文	页 数	英 文	中 文	页 数
feast on	享用	29	flower(s)	花	10
feed	喂养	33	fog(s)	雾	57
feed on	以……为食	33	Follow	跟随	20
feel	感觉, 印象	35	food	食物, 食品	8
female	雌性的	24	forest(s)	森林	2
fern(s)	蕨类植物	33	forester(s)	护林员	75
fertile	受精	24	forever	永远	32
fiddler crab	招潮蟹	58	forget	忘记	28
Fig	无花果树	6	form	形成	12
fill(s)	充满, 填	72	fox(es)	狐狸	53
filter	过滤	58	freeze(s)	结冰	8
fin(s)	鳍	59	fresh	新鲜的	65
finally	最终	31	fresh water	淡水	58
find(ing)	找, 发觉	7	friend(s)	朋友	35
fine	精细的	15	frog	青蛙	55
finger(s)	手指	42	front	前面的	56
fire(s)	火	63	Frosty	结霜的, 严寒的	21
Fireproof	耐火的	63	frozen	冻结的	62
Fireweed	火草	56	fruit(s)	水果, 果实	10
firmly	牢固地	14	full	完整的, 完全的, 充满的	20
First	最早的, 第一的	20	fume(s)	烟气	73
Firstly	首先	14	fungal	真菌的	38
fish	鱼	51	fungus (fungi)	真菌	17
fizzy	起泡的	67	furniture	家具	68
flake(s)	薄片	21	further	更远, 进一步	11
flat	平坦的, 扁平的	10	fussy	挑剔的	51
flavour	香味, 给……调味	67	gap	空隙	33
flesh	果肉	29	garden(s)	花园	5
float	浮动, 漂浮	7	gas	气体	8
flood(s)	洪水	73	General	将军	48
floor	地面	55	Gentle	温和的	57
flourish	兴旺	33	ghostly	幽灵的	60
flow	流动	43	giant(s)	巨人, 巨大的	5

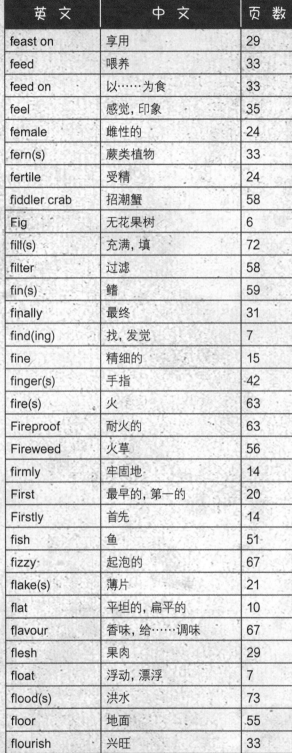

英 文	中 文	页 数
Giant anteater	巨型食蚁兽	55
giant sequoia	巨杉	48
gift	赠送	39
Giraffe(s)	长颈鹿	46
give	给	67
give off	释放	9
glide	滑行,溜	57
gloom	幽暗的	58
gnaw	啃	56
gold	黄金	69
Golden dart frog	金镖蛙	55
Golden larch	金钱松	6
good	好的	23
good at	善于,擅长	18
goodness	养分	63
goods	产品	68
gradually	逐渐地	47
grain(s)	颗粒	25
grape	葡萄	57
grass	草	57
grassy	草多的	63
great	大的	52
great grey owl	乌林鸮	60
green	绿色,绿色的	9
grey	灰色,灰色的	56
ground	地面	12
Ground beetle	步行虫	33
Group	集体,群	47
grow	生长	5
grown	成熟的,长大的	40
growth	生长	31
grub(s)	幼虫	33

英 文	中 文	页 数
guitar(s)	吉他	69
gum tree(s)	桉树	7
gust(s)	阵风	62
habitat(s)	栖息地	51
hair(s)	茸毛	15
hairy	毛茸茸的	25
hammer(ing)	敲	67
Handkerchief tree	珙桐树	6
handy	便利的	28
happen	发生	30
hard	努力地,硬的,困难的	5
hard(er)	更难的	73
Hard-wearing	耐磨的	69
hardwood(s)	硬木	69
hardy	有耐力的	62
harlequin	小丑	55
harm(ing)	危害,伤害	65
harsh	严酷的	61
Harvest	收获	66
hatch	孵化	33
Hazel	榛子	26
head(s)	头	54
headache	头痛	67
Heads up	当心,提醒	47
healthy	健康的	23
hear	听	42
hearing	听觉	42
heartwood	心材	16
heat	热,变热	62
heavy(heaviest)	最重的,最大量的	48
heavy	重的,大量的	61
height	高度,高地	30

英 文	中 文	页 数	英 文	中 文	页 数
land	陆地, 土地, 着陆	7	log	原木	33
lane	车道	43	lodgepolpine(s)	美国黑松	44
larch	落叶松	6	logger(s)	伐木工	72
large	大的	7	loner(s)	独来独往的人	43
large(largest)	最大的	48	long	长的	7
late	晚的	20	long(er)	比较长的	41
lay	产卵	33	long(est)	最长的	20
layer	层	13	look after	照顾	35
leaf	叶, 叶子	5	look out	照料	36
leafy	多叶的	10	look	看	18
learn(ing)	学习	35	loose	疏松的	74
leave(leaving)	留下	62	lose	脱离, 失去	9
leg(s)	腿	55	Lost	失去的	72
lemon(s)	柠檬	26	love	喜欢	20
level(s)	层次, 水平面	55	low(er)	下部的, 较低的	47
lichen(s)	苔藓, 地衣	17	luckily	幸运地	10
lie(s)	躺	62	lurk(s)	潜伏	58
life(lives)	生活, 生命	2	lush	郁郁葱葱的	6
lift	提高	58	made up of	由……组成	7
light	光, 阳光	9	magic	神奇	封面
like	像……那样, 喜欢	11	Mahogany	桃花心木	7
limb(s)	四肢	55	main	主要	7
Lime	酸橙树	6	mainly	主要地	66
Linden	椴树	6	make	制造, 产生	8
link	连接	38	malaria	疟疾	67
liquid	液体	9	male	雄性的	24
litre(s)	升	15	mammal(s)	哺乳动物	51
little	一点, 小的	16	Mango tree	杧果树	7
live	生活, 生存	5	Mangrove	红树	7
living	生活, 居住	2	many	很多	6
load(ed)	装载	72	maple	枫 (树)	6
Loblolly pine	火炬松	7	maple syrup	枫糖浆	6
local	本地的	2	marten(s)	貂	61

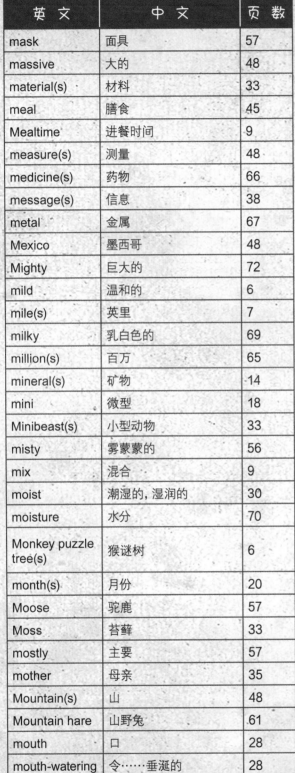

英 文	中 文	页数
northern	北方	7
Norway spruce	挪威云杉	7
nowhere	无处	72
nut(s)	坚果	21
Nutmeg tree	肉豆蔻	7
nutrient(s)	养分	29
nutritious	有营养的	33
Oak	橡树	7
object(s)	物体, 物件	68
ocean(s)	海洋	27
offer	提供	43
often	经常, 常常	2
Oil spill	漏油	15
old	老的	5
old(er)	年长的, 年老的	31
old(est)	最老的	5
Olive	橄榄	7
ooze	渗出	69
open	打开	12
orange	橘黄色, 橘黄色的	19
orchard(s)	果园	66
Oregon	俄勒冈州	57
Oregon grape	俄勒冈葡萄	57
originally	最初	67
Ouch	哎哟	44
outdoor(s)	户外	18
outer	外面的	16
outside	外面	42
outward(s)	向外的	11
Oval	椭圆的	11
overlap(ping)	重叠	60
overlook(ing)	忽略	2
owl(s)	猫头鹰	52

英 文	中 文	页数
own	自己的	30
oxygen	氧气	9
Pacific	太平洋的	56
pack(ed)	密集的, 填充的	11
page(s)	页	2
pain	痛	35
palm	棕榈树	6
pancake(s)	煎饼	67
panel	板	18
paper	纸	17
Paper birch	纸皮桦	38
papery	纸质的	26
parent(s)	父母	28
park	公园	74
part	部分, 作用	5
partner(s)	伙伴	35
party	派对	69
pass(ed)	经过, 通过, 传递	20
patch	小块土地	54
peach(es)	桃	26
peanut butter	花生酱	8
Pear(s)	梨	66
peel(s)	脱落	17
pencil(s)	铅笔	68
people	人们	52
perfect	完美的	28
period(s)	时期	62
person	个人	70
pest(s)	害虫	75
phloem	韧皮部	16
photosynthesis	光合作用	8
pick up	接收	47
piece	一片, 一块	17

英 文	中 文	页 数	英 文	中 文	页 数
pinboard(s)	记事板，针板	69	pouch(es)	袋	57
pincer	钳子	60	pounce	暴起突袭	56
Pine	松树	7	power plant(s)	发电厂	73
Pine marten(s)	松貂	61	Powerful	强大的	62
Pinecone	松果	10	precious	珍贵的	72
pip(s)	籽	26	predator(s)	捕食者	53
pipeline(s)	管道	18	prefer	更喜欢	62
Pitcher	瓶状叶	55	prepare	准备	28
pizza	比萨饼	66	prey	猎物	58
place(s)	地方，放置	6	print(s)	拓印	17
place mat(s)	餐垫	69	prise(s)	撬开	60
plain(s)	平原	63	Proboscis monkey	长鼻猴	58
planet	行星	48	process	过程	8
plant	植物	5	produce(s)	产生	6
plantation(s)	种植园	66	product(s)	产品	65
play	起作用	70	promise	承诺	23
plenty	充足的	57	pronounce(d)	发音	38
plop	扑通	27	protect(s)	保护	17
pluck	采摘	46	protective	保护的	63
plum(s)	李子	10	proud	骄傲的	75
pointy	尖的	11	provide	提供	16
poke(s)	刺，捅，戳	30	prowl(s)	潜行	59
pollen	花粉	24	pulp	纸浆	68
pollute	污染	73	Puma(s)	美洲狮	56
pollution	污染	15	pump	注入	45
Polylepis tree(s)	多鳞树	49	pumper(s)	泵	18
pomegranate(s)	石榴	66	Purple	紫色，紫色的	55
pond(s)	池塘	56	push	推	15
Ponderosa pine tree(s)	西黄松	63	put(putting)	放	28
poo	粪便	29	puzzle	谜	6
Poplar	白杨	7	Pygmy marmoset	侏儒狨猴	55
possible	可能	14			

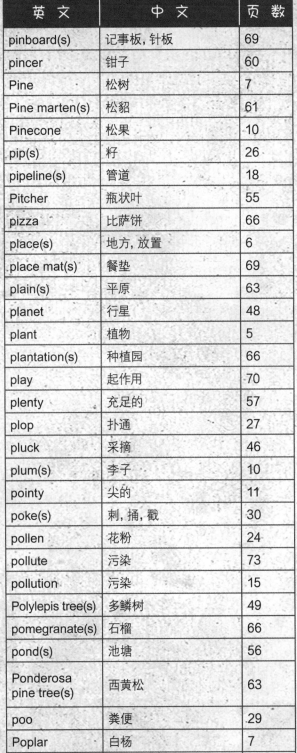

英 文	中 文	页 数	英 文	中 文	页 数
start	开始	20	sugar maple	糖枫	67
stay	保持	8	sugary	含糖的	9
steady	稳定的	13	Summer	夏天, 夏季	9
steal	偷	57	sun	太阳	9
Stealthy	诡秘的	55	Sundarbans	孙德尔本斯	59
steamy	潮湿的	54	sunlight	太阳光	8
steep	陡峭的	51	sunny	阳光照耀的	62
step(s) in	介入	40	sunshine	阳光	20
stick	粘	25	super	超级的	53
sticky	黏的	44	supermarket	超市	52
still	仍然	23	support(s)	支撑, 支持	13
stilt(s)	高跷	58	surely	无疑地	72
stone	核	26	surface	表面	35
stony	多石的	51	surround(ed)	围绕	16
stop(s)	阻止, 停止	17	surrounding(s)	周围	56
store	贮存	28	survive	生存	5
storm	风暴	14	survivor(s)	幸存者	51
story	故事	32	Swamp	沼泽	58
straight	直的, 直接的	14	swap	交换	38
strange	奇怪的	59	sweet	甜的	25
Strangler fig	绞杀榕	7	swell(s)	胀	30
stream(s)	溪流	56	swing	荡来荡去	55
street(s)	街道	51	Sycamore	美国梧桐	27
stretch(es)	延伸	7	syrup	糖浆	6
stripe	条纹	57	system	系统	28
striped	有条纹的	59	tactic	战术	45
strong	强壮的, 坚固的	13	taiga	泰加林	60
strongest	最强壮的	40	tail(s)	尾巴	55
struggling	挣扎	40	take care of	照顾	40
stump	树桩	13	take part in	参与	43
sturdy	壮实的	13	talk(ing)	谈话	43
suck	吮吸	15	tall	高的, 高大的	11
sugar	糖	16	taller	更高的	5

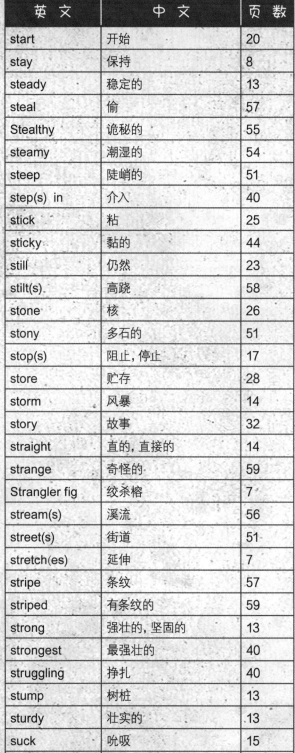

注：以上中英对照的词义只限于本书内容范围。

致　谢

The publisher would like to thank the following people for their assistance:
Hélène Hilton, Jolyon Goddard, Katie Lawrence, Clare Lloyd, and Abigail Luscombe for editorial help; Kitty Glavin and Eleanor Bates for design help; Helen Peters for the index; Akash Jain for picture research; and Tom Morse for CTS help. Many thanks to Rae Spencer-Jones and Simon Maughan at the RHS.

PICTURE CREDITS

The publisher would like to thank the following for their kind permission to reproduce their photographs: (Key: a-above; b-below/bottom; c-centre; f-far; l-left; r-right; t-top)

1 123RF.com: alein (cl); Eric Isselee (bc). Dorling Kindersley: Paradise Park, Cornwall (tr). 2-3 Dreamstime.com: Designprintck. 4-5 Alamy Stock Photo: Carolyn Clarke. 5 Dreamstime.com: Designprintck. 6 123RF.com: Maksym Bondarchuk (cla, c); joseelias (cb/Rubber tree); 1ธ Jaturon Ruaysoongnern (cb/Mahogany Tree, cb, bc); fotoplanner (cla/Young birches); Smileus (fbl). Dreamstime.com: Pablo Caridad / Elnavegante (tr). iStockphoto.com: tiler84 (bc/Fig tree). 6-7 Dreamstime.com: Designprintck (b/Background); Ruslan Nassyrov / Ruslanchik (Background). 7 123RF.com: Maksym Bondarchuk (c); Valentyn Volkov (fcl); Cherdchai Chaivimol (ca/Bodhi Tree); marigranula (c/Date palm); liligraphie (ca); fotoplanner (ca); 1ธ Jaturon Ruaysoongnern (crb, fbl). Dorling Kindersley: Lindsey Stock and Lindsey Stock (cra). Dreamstime.com: Lano Angelo / Dina83 (fcl/Baobab tree). iStockphoto.com: tiler84 (cla/Common Fig tree). 8 123RF.com: fotoplanner (bc). Dreamstime.com: Alexander Potapov (br). 9 123RF.com: belchonock (br). Dorling Kindersley: (bl); Ian Cuppleditch (t). Dreamstime.com: Alexander Potapov (bc). 10 123RF.com: liligraphie (fcl). Dorling Kindersley: E. J. Peiker (c). Dreamstime.com: Mikelane45 (cr). 11 Dreamstime.com: Designprintck (Background). 12 123RF.com: alein (ca/Woodpecker); Roman Iegoshyn (tr); Eric Isselee / isselee (cra). Dorling Kindersley: Alan Murphy (cb); RHS Wisley (Crabapple); E. J. Peiker (cla, cb/Owl); Paradise Park, Cornwall (ca/Bluebird). Dreamstime.com: Mikelane45 (ca). 13 123RF.com: wklzzz (cl). 14 Dorling Kindersley: Paolo Mazzei (ca). 15 Dorling Kindersley: Barrie Watts (ca); Stephen Oliver (fcla). 16 Getty Images: Don Mason (b). 17 Alamy Stock Photo: Rolf Nussbaumer Photography (c); Snap Decision (l); Colin Varndell (r). Dreamstime.com: Designprintck (Paper). 19 123RF.com: Christian Mueringer. Dreamstime.com: Designprintck (Background). 20 123RF.com: Agata Gladykowska (cla); Roman Iegoshyn (bc). Dorling Kindersley: Batsford Garden Centre and Arboretum (bl); Natural History Museum, London (fcr). Dreamstime.com: Jessamine (ca). 21 123RF.com: bmf2218 (tl, tc, cla). Alamy Stock Photo: WILDLIFE GmbH (clb/Norway Maple Leaf). Dorling Kindersley: Jerry Young (cb). Dreamstime.com: Motorolka (clb). Fotolia: Eric Isselee (cl). PunchStock: Corbis (fcl). 22-23 123RF.com: ammit. 23 Dreamstime.com: Designprintck (Background). 24 Dorling Kindersley: Alan Buckingham (cb). 25 Dorling Kindersley: Alan Buckingham (cr); Jerry Young (ca). 26 Dreamstime.com: Tamara Kulikova / Tamara_k (fbl); Alex Bramwell / Spanishalex (tr). 27 Dreamstime.com: M Schaefer (ca). 28 123RF.com: Dule964 (b/Leaves); Isselee (bc). Fotolia:

Steve Byland (cr). Getty Images: Paul E Tessier / Photodisc (clb). 29 123RF.com: Eric Isselee (cb, cra). Alamy Stock Photo: Duncan McKay (cl). 30 123RF.com: madllen (cb). Dreamstime.com: 3drenderings (r). 31 123RF.com: alein (ca). Dreamstime.com: Dule964 (Leaves). 32 Dreamstime.com: Sergei Razvodovskij / Snr (cla). 33 123RF.com: Oksana Tkachuk (cra); wklzzz (br). Dorling Kindersley: Paolo Mazzei (cb). Dreamstime.com: Isselee (c). 34-35 Dreamstime.com: Inga Nielsen / Inganielsen (b). 35 Dreamstime.com: Designprintck (Background). 36 Dorling Kindersley: Stephen Oliver (b). 40 123RF.com: neydt (bl). 41 123RF.com: avtg (clb). 42-43 123RF.com: Ralph Schmaelter (Davidia involucrata). Dreamstime.com: Designprintck (Paper). 44 Alamy Stock Photo: Historic Collection (clb); Universal Images Group North America LLC / DeAgostini (l). Dreamstime.com: Alexander Potapov (tc). 45 Alamy Stock Photo: Steven Prorak (cl). 46-47 iStockphoto.com: Orbon Alija. 48-49 Dreamstime.com: Designprintck (b). 50-51 Alamy Stock Photo: Mint Images Limited. 51 Dreamstime.com: Designprintck (Background). 52 Dorling Kindersley: Kim Taylor (cla); Jerry Young (cra); Paolo Mazzei (bc). 52-53 123RF.com: Andrzej Tokarski / ajt (cb). Dreamstime.com: Dule964 (b/Leaves). 53 123RF.com: Steve Byland / steve_byland (cla/Sapsucker); Eric Isselee / isselee (ca); wklzzz (cl). Dorling Kindersley: British Wildlife Centre, Surrey, UK (cr); Natural History Museum, London (cla); Paolo Mazzei (clb). Dreamstime.com: Alle (cla/Bee); Jessamine (ca/Nest); Jim Cumming (cb). 54 Dorling Kindersley: Natural History Museum, London (cla); Andy and Gill Swash (clb). SuperStock: Glenn Bartley / All Canada Photos (cra). 55 123RF.com: Michael Zysman / deserttrends (cla). Alamy Stock Photo: Amazon-Images (c/Ant); Andrew Barker (ca); Life on White (br). Dorling Kindersley: Thomas Marent (ca/Pitcher Plant); Jerry Young (fcl, c/Leopard); Natural History Museum, London (fbr). Dreamstime.com: Eric Isselee (c/Marmoset, cb); Pablo Hidalgo / Pxhidalgo (tc); Matthijs Kuijpers (cra). Getty Images: Martin Harvey / Photodisc (c). 56 Alamy Stock Photo: imageBROKER (c). Dreamstime.com: Philip Bird (crb); Vivian Mcaleavey (cla); William Bode (clb); Musat Christian / Musat (cb); Jnjhuz (clb/Beaver). 57 123RF.com: Mariusz Jurgielewicz (br). Alamy Stock Photo: imageBROKER (fclb, clb). Dorling Kindersley: Booth Museum of Natural History, Brighton (cla). Dreamstime.com: Gunold Brunbauer / Gunold (cr). 58 Dorling Kindersley: Greg and Yvonne Dean (tc). Dreamstime.com: Feathercollector (cr); Trubavin (cra); Mikhail Blajenov / Starper (clb). 59 Dorling Kindersley: Greg and Yvonne Dean (ca); Jerry Young (cb); Andy and Gill Swash (tl). Dreamstime.com: Chatchawin Pola / Chatchawin (clb); Suradech (c). 60 123RF.com: Maksym Bondarchuk (cla, cra); zerbor (cra/Silver birch). Dreamstime.com: Gunold Brunbauer / Gunold (fcrb); Josefpittner (tc); Steve Byland / Stevebyland (clb); Vanessa Gifford / Vanessagifford (crb). 61 123RF.com: belchonock (c/Tree); zerbor (cla, ca/Silver birch, fclb/Silver birch); Maksym Bondarchuk (ca, cra, cb); liligraphie (c); fotoplanner (cr, c/Fir tree); mediagram (c/Pine trees). Dreamstime.com: Helen Panphilova / Gazprom (clb); Horia Vlad Bogdan / Horiabogdan (fclb); Moose Henderson / Visceralimage (clb/Pine Martine); Scattoselvaggio (bc). 62 Alamy Stock Photo: imageBROKER (clb). 63 Fotolia: Eric Isselee (cl, fcl); Steve Lovegrove (clb). 64-65 123RF.com: ronstik. 65 Dreamstime.com: Designprintck (Background). 66 Dreamstime.com: Grafner (tl); Valentyn75 (crb). 67 123RF.com: Akulamatiau (clb); sunteya (cl); Gabor Havasi (tc); Karandaev (ca); Malosee Dolo (cb/Bottle). Dreamstime.com: Isabel Poulin (bl); Roman Samokhin (cb). 68 123RF.com: Katarzyna Białasiewicz (fcr/Table chair); serezniy (cra); Turgay Koca (fcr); Sakarin Plangson (ca); wklzzz (l). Dreamstime.com: Pictac (cr). 68-69 123RF.com: wklzzz (bc). 69 123RF.com: George

Tsartsianidis (tc); wklzzz (r). Dreamstime.com: Dmitry Rukhlenko / F9photos (bc); Piotr Adamowicz / Simpson333 (cla). Fotolia: sisna (bl). 70 Dreamstime.com: Ruslan Nassyrov / Ruslanchik (bl). 72-73 Dreamstime.com: Designprintck (Background). 74-75 Dreamstime.com: Designprintck (Background). 75 Dreamstime.com: Andrzej Tokarski (c). 76-77 Dreamstime.com: Designprintck (Background). 76 123RF.com: joseelias (bc); Valentyn Volkov (tc). 77 123RF.com: fotoplanner (tc). 78-79 Dreamstime.com: Designprintck (Background). 80 Dreamstime.com: Designprintck (Background)

Cover images: Front: 123RF.com: Eric Isselee / Isselee clb,c, Yuliia Sonsedska / sonsedskaya cb/ (Raccoon); Dorling Kindersley: British Wildlife Centre, Surrey, UK cb, crb, RHS Wisley; Back: 123RF.com: Eric Isselee / Isselee c; Dorling Kindersley: Paradise Park, Cornwall cla, RHS Wisley; Fotolia: Eric Isselee cl

All other images © Dorling Kindersley
For further information see: www.dkimages.com

ABOUT THE ILLUSTRATOR

Claire McElfatrick is a freelance artist. She created illustrated greetings cards before working on children's books. Her hand-drawn and collaged illustrations for *The Magic & Mystery of Trees* are inspired by her home in rural England.

Thank you to Mary Ling, who grew this book from a seed.

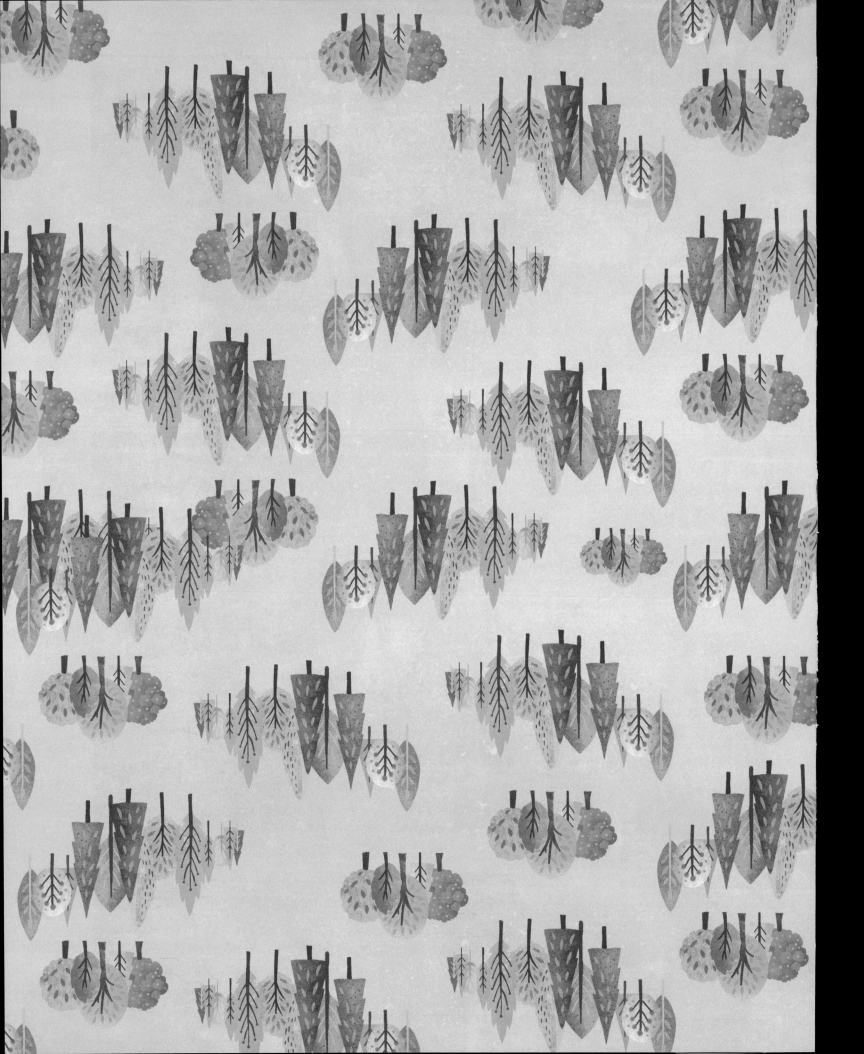

THE MAGIC & MYSTERY OF Trees

DK

From the deepest, densest forests
to our local towns and cities,
trees
are all around us.

We share our
world with trees, living side by
side with them but often
overlooking them.

Wander through the pages of
this book to discover the secret
lives of trees.

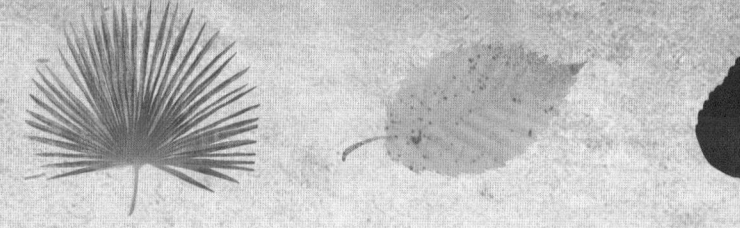

CONTENTS

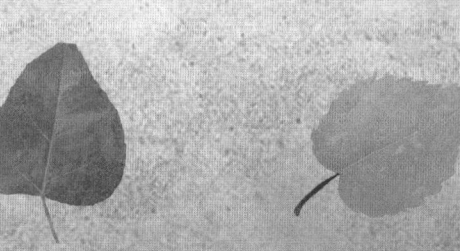

WHAT IS A TREE ?

A tree is a huge plant that towers above us. You'll find trees standing alone in back gardens or clustered together in thick forests.

Trees are true wonders of nature. Some species can grow taller than a stack of 50 cars one on top of the other! Trees can live for hundreds of years, and the very oldest are thousands of years old.

Every part of a tree works together. From the deepest roots that burrow through the earth to the smallest leaf on the highest branch, every part of a tree is working hard to help it survive.

When you get to know these silent giants, you'll never look at trees the same way again...

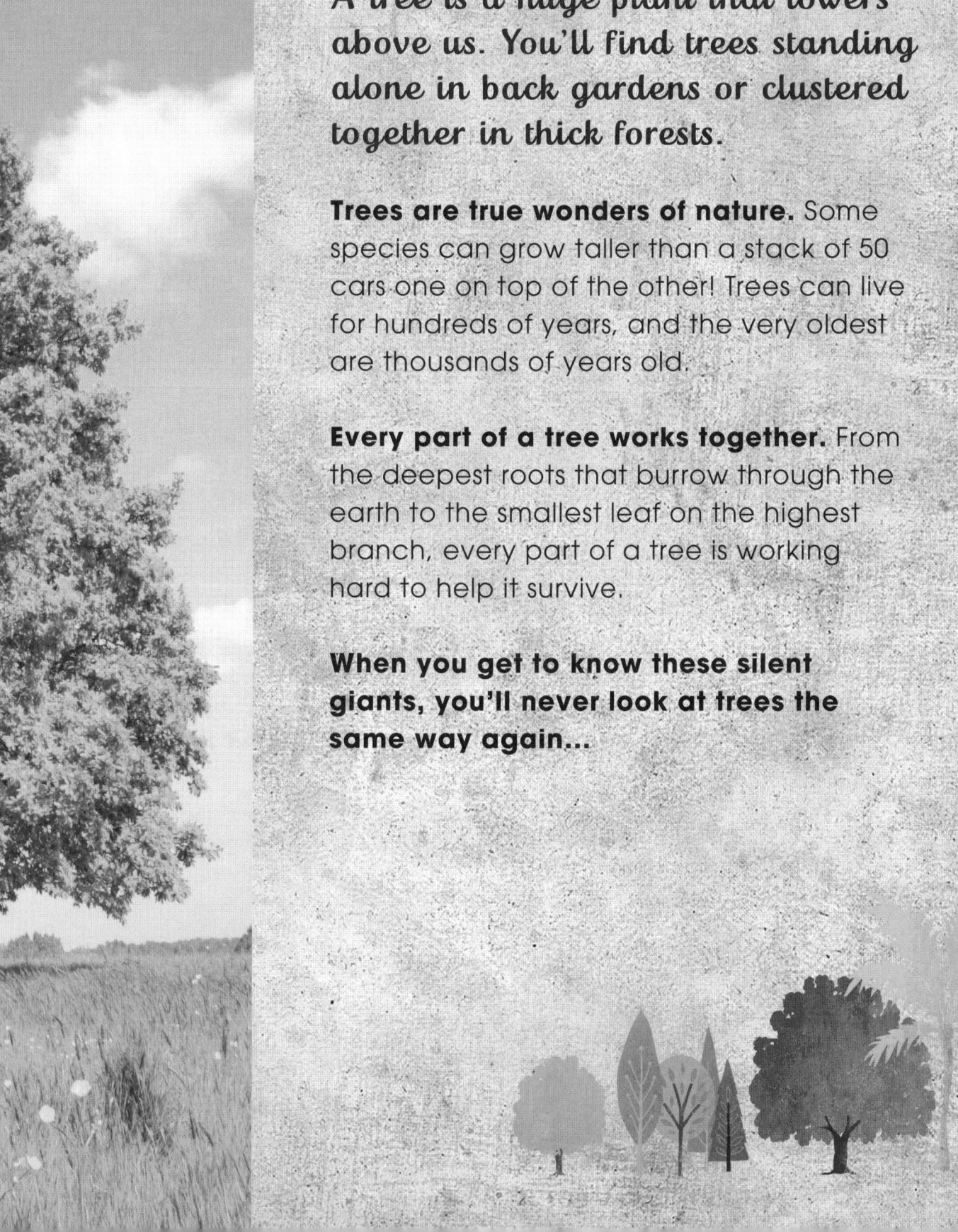

Where in the world?

From rocky coasts to lush valleys, trees are found almost everywhere. Forests are places where many trees grow together.

Canada's most famous tree, the **maple**, produces maple syrup.

The world's tallest trees, **redwoods**, live in western North America.

North America

Broadleaved forests grow in parts of North America and Europe with mild climates.

South America

The biggest **rainforest** in the world is the Amazon in South America. **Tropical rainforests** grow close to the Equator, around Earth's middle, where it is very hot all year round.

Monkey puzzle trees grow in Chile, at the tip of South America.

N
NW NE
W E
SW SE
S

KEY

Acacia
Ash
Aspen
Banyan
Baobab
Birch
Cabbage tree
Cedar
Chinese fire tree
Cocoa
Coconut palm
Date palm
Douglas fir
Elm
Eucalyptus
Fig
Golden larch
Handkerchief tree
Huasai palm
Jacaranda
Japanese beech
Juniper
Kapok
Kauri
Lime
Linden

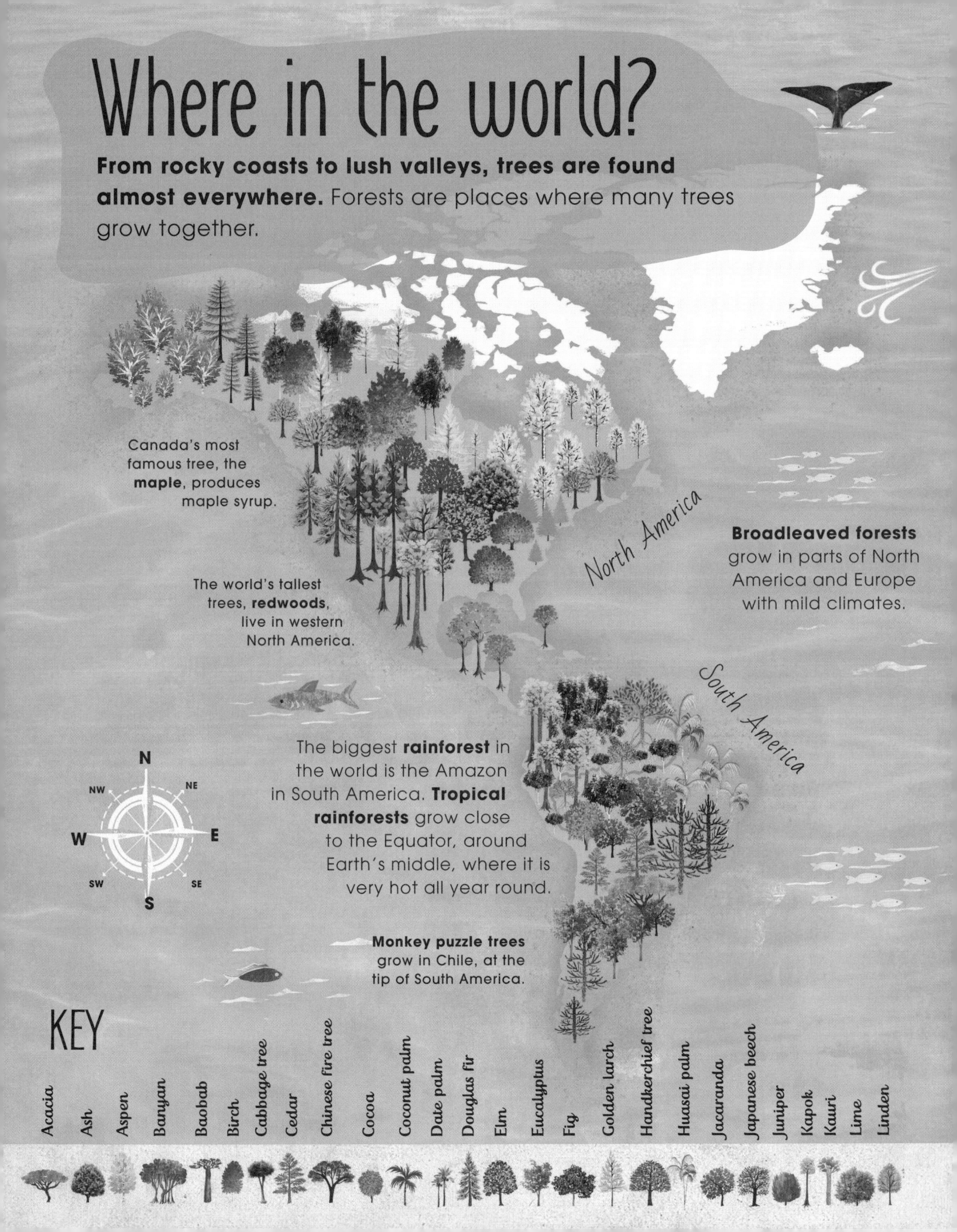

FOREST TYPES

There are three main types of forests: **broadleaved forests**, **conifer forests**, and **rainforests**. Each of these forests is made up of different types of trees.

A huge **conifer forest** stretches across northern North America, Russia, and northern Europe. These places have long, snowy winters.

Europe

Asia

Africa

Coconut seeds can float for miles before finding a place to grow.

Kauri pines live only in New Zealand. They can get very old and large.

Eucalyptus, or gum, trees grow in Australia's dry forests. They keep their leaves all year round.

Australia

Forests cover almost a third of Earth's dry land.

Loblolly pine

Mahogany

Mango

Mangrove

Monkey puzzle

Maple

Norway spruce

Nutmeg

Oak

Olive

Pine

Poplar

Quaking aspen

Ramon

Redwood

Red maple

Rowan

Rubber

Sacred fig.

Sausage tree

Silver birch

Spruce

Sitka spruce

Strangler fig.

Teak

Umbrella thorn

How trees live

You've never seen a tree eat a bowl of noodles or a peanut butter sandwich, so what do they eat? As long as it has sunlight, water, and a gas called carbon dioxide, a tree can live, grow, and even make its own food!

The amazing food-making process of plants is called photosynthesis.

Trees are tough, but they must stay warm to survive. If the water in the tree's leaves freezes, it can't make food for itself.

Mealtime

A tree's green leaves soak up light from the sun. Then they use energy from the light to mix carbon dioxide and water. This makes a sugary liquid called sap, which is the tree's food.

Making oxygen

While they are busy making sap, the tree's leaves give off a gas called oxygen. All animals, including us, breathe in oxygen and breathe out carbon dioxide. If there were no plants such as trees, we wouldn't have air to breathe.

Summer days

Broadleaved trees only make food in spring and summer, because there is more sunlight. They lose their leaves in the autumn. Conifers can have leaves or needles. They keep making food throughout the winter.

Types of trees

With so many trees growing around the world, it can be tricky to tell one leafy plant from another. Luckily, there are only two main families of trees: **broadleaved** trees and **conifers**.

Oak leaves

Pinecone

Needles

Broadleaved

These trees have wide, flat leaves. They all make flowers, although some are almost too small to see. Their seeds ripen inside juicy fruits, such as plums or figs. Most broadleaved trees drop their leaves in autumn and grow new ones in spring.

Conifer

These trees have long, thin leaves called needles. Many conifers are called evergreens because they keep their leaves all year round. Conifer seeds are found inside hard, knobbly cones, such as pinecones.

ROUND

Oak

The branches of a round tree spread themselves evenly upwards and outwards from the trunk.

BROAD

Maple

A broad tree has branches that spread further to the sides than upwards.

SPREADING

Banyan

These trees have branches that grow up and out to create a spreading shape.

OVAL

Hornbeam

Oval trees have a rounded shape that is taller than it is wide.

The leafy part of the tree above the trunk is called the crown. It comes in different shapes. Many broadleaved trees are wide and round, while conifers are often shaped like cones.

WEEPING

Willow

A weeping tree has branches that droop downwards.

PALM

Coconut palm

Palms are broadleaved trees that grow in hot countries. Unlike other trees, they don't grow side branches.

TALL AND THIN

Cypress

Some trees have closely packed branches that grow upwards. Many conifers are tall and thin.

CONE

Spruce

A cone-shaped tree's branches get shorter as they go up the the trunk, ending in a pointy tip.

Parts of a tree

BRANCHES

Branches grow from the trunk. They divide to form smaller branches, which end in twigs. Leaves sprout from twigs. Flowers and fruit grow from twigs at certain times of year.

CANOPY

High above the ground, twigs and leaves weave together to form a dense, dark blanket called the canopy.

Bud

In spring, buds burst open and leaves and flowers unfurl from them.

STUMP

When a tree is cut down or its trunk breaks, it leaves behind a stump.

Bark is a thin, tough layer that covers the trunk.

TRUNK

The sturdy trunk grows from the ground. It is very strong, and supports the weight of the tree's branches.

Wherever they grow, all trees have the same parts: roots, a trunk, branches, and leaves.

ROOTS

Underground roots hold the tree steady in the ground.

Secret roots

In the damp, dark world below, roots spread through the soil to form a woody network. Up to a third of the tree is hidden underground.

Cosy homes
Rabbits and tiny creatures such as worms and beetles live among the roots.

Holding fast
Roots have two main jobs. Firstly, they hold the tree firmly in the ground, so it will not blow over in a storm. Secondly, they draw up water containing minerals from the soil, so the leaves can make food.

Reaching out
Some trees have a big main root called a **tap root**. While most roots grow sideways, the tap root shoots straight down. A tree's root network often spreads wider than the tree is tall, to find as much water as possible.

Trees do most things slowly, but they drink very fast! A big tree can suck in hundreds of litres of water from the soil every day.

Trunk

Pollution
Tree roots are very sensitive. They can sense pollution in the soil and avoid it by growing in a different direction.

Oil spill

Water seekers
The main roots divide into smaller ones. The smallest ones at the end of the root are called **rootlets**. They are covered in fine hairs that can sense water.

Water travels **up** the **roots**.

Rootlets

Big and tough
The main roots are strong and woody, like branches. Each root tip has a tough cap to push through the soil as it grows. These big roots spread up to 1.5 m (5 feet) into the ground.

Trunk and bark

A tree's trunk supports its branches, just like your skeleton holds up your body. The trunk has to be very sturdy to support the huge weight of all the branches. **A tree simply wouldn't be a tree without a trunk!**

Inside the trunk

At the centre of the trunk is the heartwood. This grew when the tree was young. It is surrounded by sapwood, which contains tiny tubes that carry water from the roots to the leaves.

Between the sapwood and the outer bark is a very thin layer called the phloem. This carries sugar from the leaves to the rest of the tree.

Phloem

Bark

Tree rings provide clues about the tree's history. **Wide rings** show years when the tree grew quickly. **Narrow rings** show when the tree grew only a little, because conditions were too cold or dry.

Heartwood

Sapwood

Bark, like this peeling birch bark, is the outer layer of the trunk. It stops the tree drying out, and also protects it from insects and fungi.

Young trees have **smooth bark**. As trees get older, their bark cracks, peels, and becomes **more wrinkly**, like this scaly tree.

Try making **bark prints** by rubbing crayons onto a piece of paper placed on bark. The **texture** will come through.

Different types of trees have different bark. They are also home to **lichens**, like these blotchy yellow ones.

Water pumpers

Veins are like tiny pipelines running through the leaf. They take in water from the tubes in the trunk's sapwood, and carry food made by the leaves to the rest of the tree.

Leaves

Next time you are outdoors, take a close look at a leaf. Leaves are very special, as it's in the leaves that the tree works its magic by making its own food.

Beech leaf

Veins

Light catchers

Broadleaved trees spread their wide, flat leaves to capture as much light as possible. Each leaf is like a mini solar panel, soaking up energy from the Sun.

Leaf shapes

Each tree has leaves with a slightly different shape. They can be long and thin, or wide and round. Flat, round leaves are good at catching sunlight, but also lose more water.

Trees can't move from place to place, but they can very slowly turn their leaves to face the sun.

Green
summer leaf

Fading
to yellow

Dry,
brown leaf

Leaves are green
because they contain
a natural colour
called chlorophyll.
In autumn the green
fades, and other
colours in the leaves
can be seen. They
turn yellow, orange,
and brown.

Turning orange
in autumn

19

Follow one maple tree through the seasons...

SPRING

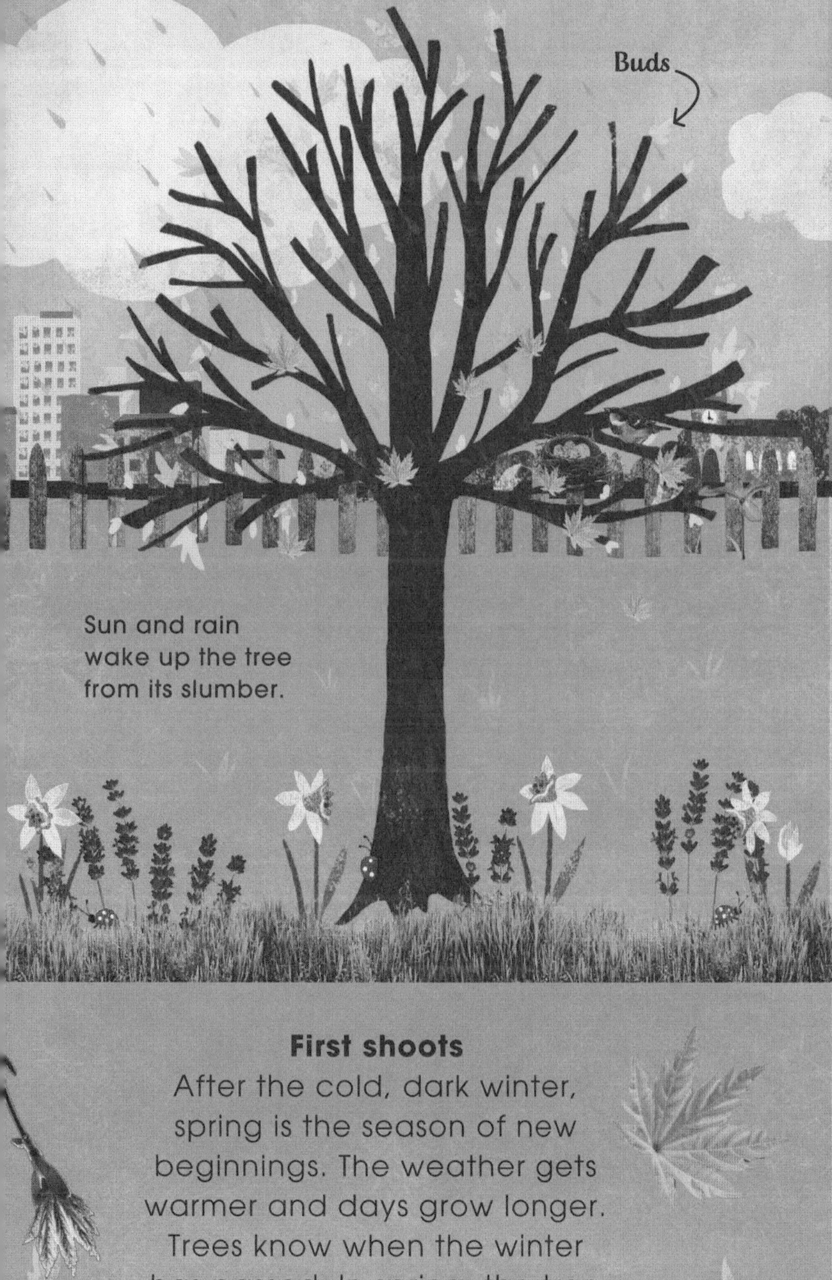
Buds
Sun and rain wake up the tree from its slumber.

SUMMER

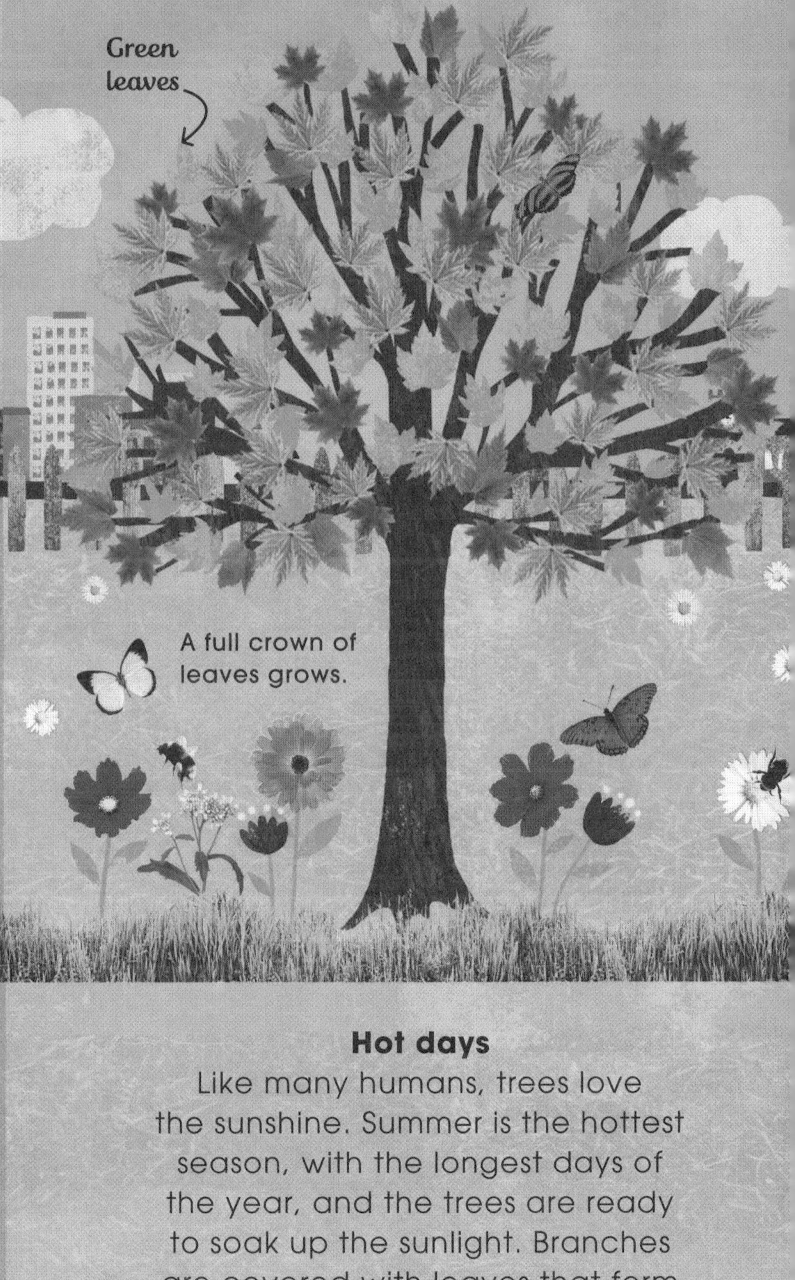
Green leaves
A full crown of leaves grows.

First shoots

After the cold, dark winter, spring is the season of new beginnings. The weather gets warmer and days grow longer. Trees know when the winter has passed. In spring, the tree grows green buds. The buds open, uncurling and spreading new leaves in the sunshine. The tree's flowers bloom.

Hot days

Like many humans, trees love the sunshine. Summer is the hottest season, with the longest days of the year, and the trees are ready to soak up the sunlight. Branches are covered with leaves that form a dense, shady layer. In late summer, the tree's fruits start to grow. Trees do most of their growing during the summer months.

AUTUMN

Red leaves

Leaves lose their green colour and fall.

Changing colours
In autumn, the weather turns cooler and days get shorter. Fruits and nuts ripen, and trees spread their seeds. Flat, wide leaves can catch blustery autumn winds, so the tree sheds them to avoid damage to its branches. Its green leaves turn yellow, orange, red, and brown, then drift down to the ground.

WINTER

No leaves

Bare branches won't get weighed down with snow.

Frosty flakes
Winter is the coldest season, with the shortest days. The tree has shed its leaves. It may look dead, but it is merely sleeping as it waits for spring. The tree has moved the stocks of sap it spent all summer making away from its branches and into its roots.

FLOWERS, FRUITS, AND SEEDS

Trees need to make new trees for the forest to stay healthy. That's why they grow flowers, fruits, and seeds.

The flowers bloom in spring. During the warm days of spring, bees buzz from tree to tree, visiting as many flowers as they can.

Seeds are little parcels that contain a whole new tree, ready to grow. They ripen in summer and autumn. Each seed needs to find a good place to grow.

Trees, like all living things, are born, grow, and eventually die. But they leave behind the promise of new forests still to come...

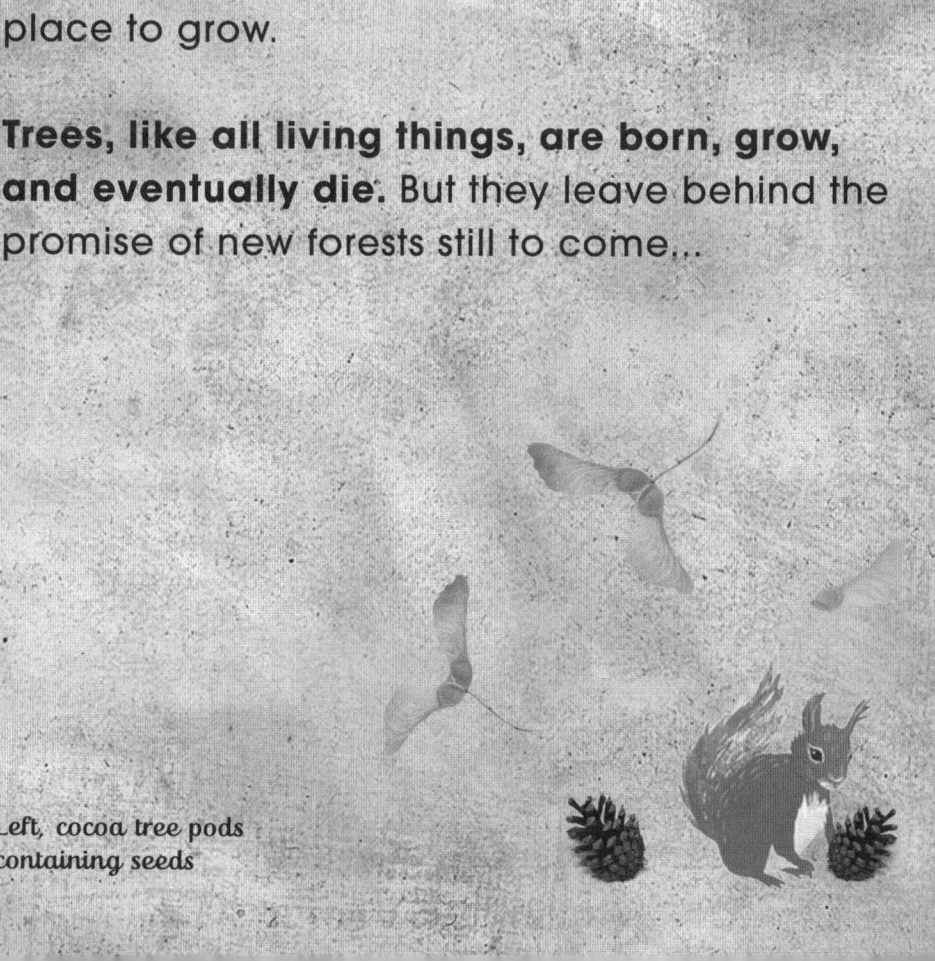

Left, cocoa tree pods containing seeds

Flowers

Trees that want to attract **insects** have large, showy flowers, while ones that spread pollen on the **wind** have tiny, delicate ones. Some tree flowers are so small that they can be tricky to spot, but they have vital work to do.

Blossom

In spring, apple and cherry trees are covered with flowers we call **blossom**. These flowers make the trees look beautiful, but their main job is to tell insects such as bees that there's a treat ready for them.

Cherry tree blossom

Did you know flowers have male and female parts? The male part produces **pollen**. The female part makes tiny **eggs**. Male pollen must join with these eggs to make them fertile and then they can ripen into seeds.

Little helpers

Insects visit flowers to drink a sweet liquid called **nectar**. Any pollen grains on its body rub off inside the next flower the bee visits. This makes the next flower fertile (able to make seeds).

Grains of pollen stick to the bee's **hairy body**.

The bright colours and sweet smells of flowers tell the insects the nectar is ready.

Pollen

Many trees, such as pine and fir, spread their pollen on the wind.

Conifer
seeds

Fruits and seeds

Once the tree's flowers have been made fertile by pollen, it is **time for seeds to grow**. They can develop inside fruit, cones, hard shells, or papery covers.

Conifer seeds

Most conifer trees make their seeds inside **cones**, not fruits. When the seeds are ripe, the cones open. The light, **papery seeds** tumble out and blow away on the breeze. Yew and juniper are unusual conifers. They make small, bitter berries. Birds love them!

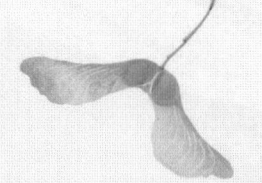

Cherries

Fruits

If you've ever enjoyed a crunchy apple or juicy cherry, you've eaten the fruit of a broadleaved tree. Mangoes, peaches, and cherries contain just one large seed, called a **stone**. Apples, oranges, and lemons have many small seeds called **pips**.

The hard shells on these seeds look very different, but they all do the same job. They **protect seeds** and help them to spread.

Acorns

Chestnut
conkers

Hazel, chestnut, and walnut trees make seeds with a hard shell – we call these **nuts**. Acorns are the seeds of the oak tree.

Sycamore seeds

Maple and sycamore seeds have a double wing. They spin like mini-helicopter blades, to land far away.

Blown on the wind

Tree seeds must be scattered far and wide so that new trees can grow. Some seeds are scattered by the wind. Sycamore, maple, and ash trees have light, **winged** seeds that spin through the air.

Carried by currents

Trees that live by rivers and oceans make seeds that float. The current carries them away. **Coconut palms** grow on the seashore in warm places. Ripe coconuts plop into the water. The tide washes them away to root on distant coasts.

Unripe, green coconuts

Coconuts are light enough to **float**.

27

Animal assistants

Animals love the bright colours and mouth-watering smells of fruit. Trees get animals to **spread their seeds** for them by putting them inside delicious things.

ADVENTURE TIME

For children, the best place to grow up is near our parents, who provide everything we need. Trees are very different – they like their seeds to travel alone to distant places. Fast-moving animals provide a perfect transport system.

Buried nuts

In autumn, **squirrels** and **jays** prepare for winter by burying nuts and acorns. Nuts make a handy food store during the long, bleak months of winter – as long as the animal doesn't forget where its food is buried! Any forgotten seeds will sprout into new trees in the spring.

Monkey dung

Monkeys love to feast on figs.
They can digest the juicy flesh,
but the hard seeds pass right
through the animals' bodies
and come out in their poo!

**As the monkeys
wander from tree
to tree**, the seeds
in their dung get
scattered all over
the forest. Monkey
poo contains all the
nutrients a seed needs
to sprout and grow
strong and healthy.

Tropical fig tree

29

From seed to tree

Trees are the tallest living things in the world but they are born from tiny seeds. Growing to their full height can take 100 years. Here's how a young oak sprouts from a little acorn, to grow taller than a house.

Seeds are full of food to keep the seedling going until it can make its own food.

A young tree is called a seedling, or sapling.

Acorn

Root

Sprouting

If a seed lands in moist soil in a warm, light place, big changes start to happen. The seed swells and the case splits open. A tiny root pokes down to take in water.

Getting taller

A small, green shoot pushes up through the soil. Once the first leaves unfurl in the sunlight, the young plant can make its own food. A new, little tree is born.

Oak trees keep getting wider for 500 to 600 years.

Growing older

Humans grow during their childhood, but we stop when we become adults. **Trees are different** – they carry on growing. What's more, they can live at least five times as long as we do. At **100 years old**, many trees are still youngsters!

Taller and wider

As the tree gets older, its upwards growth slows and finally stops when it reaches **full height**. But if there is space, its branches and trunk carry on growing wider. The trunk of a big, old tree grows about 2.5 cm (1 inch) wider each year.

Life after death

No living thing goes on forever. Trees can live for hundreds of years, but in the end even they grow old and die. Winds shake the dead tree until its trunk cracks and it comes crashing to the ground.

A new home

But that's not the end of the story. A dead tree becomes a home for thousands of small creatures that like damp, dark places. Little beasts such as slugs, worms, woodlice, centipedes, insects, and spiders move in.

Centipede

Toadstool

Earwig

Earthworm

Fly agaric mushroom

Up to a third of all creatures that live in the forest like to live in, or eat, rotten wood.

Recycling nutrients

Minibeasts and fungi feast on rotten wood and break it down. This allows the raw materials that formed the tree to return to the soil. These nutrients feed young trees and other plants, giving them a good start in life.

The fallen tree has left a gap in the forest canopy, which allows light to reach the ground. This helps seedlings to flourish. Sometimes a seedling will even sprout from a log.

Brown garden snail

Garden spider

Moss, ferns, and flowers take root in the squelchy, nutritious wood of rotten logs and tree stumps. Fungi spread their threads through the damp wood, and mushrooms sprout from the trunk.

Centipede

Common garden slug

Ground beetle

Woodlice

Beetles lay their eggs under the bark. When the grubs hatch they feed on the rotten wood.

33

TREE PARTNERS

Trees grow and change so slowly that it's hard to tell how busy they really are. Year after year, trees stand still and silent, but there's a lot going on under the surface.

Trees are mysterious things. Recently, scientists have found out that the trees in a forest work together. They make friends and support one another. Trees look after their neighbours, and mother trees pass food to their children and older trees within the family.

We now know trees have senses. They can smell, taste, touch, and feel pain, like we do. Trees can sense danger and defend themselves against enemies. **We are learning that there's a lot more to trees than we ever imagined.**

Living together

Life in the natural world is tough. It's easier for trees to survive when they help one another. Trees living in a forest grow best if all the trees are healthy. If one tree is in trouble, the others help it. Trees also work together to make the forest warmer and more sheltered in winter, and cooler, damper, and shadier in summer.

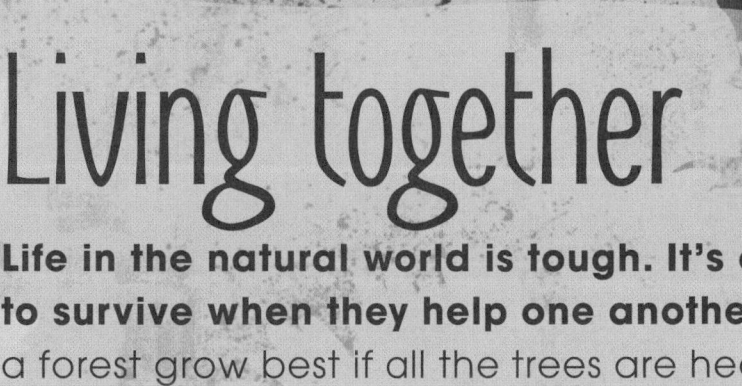

Family and friends
Trees of the same species look after one another. Oak, beech, and spruce trees only like to share water and food with their own kind. But in some places, trees of different types look out for one another too.

Stronger together

Trees grow wider until they reach the next tree. This creates a roof of branches and leaves that protects the forest from storms. If too many trees die and leave gaps, strong winds can enter and wreck the forest.

If a tree is damaged and starts to die, its neighbours will pass it food to keep it alive.

Roots

Roots spread through the soil to form a hidden web. Forest neighbours stay in touch with one another and pass food to each other through their roots.

Wood wide web

Trees like to keep in touch with each other. Experts have found out that it's not just **roots** that link the trees in a forest. They are also connected through **fungi** – the living things we know as mushrooms.

Paper birch

Hyphae

Helpful fungi like these are called **mycorrhizae** (pronounced my-cor-riz-eye).

The fungal network

Fungi are a bit like plants, but they cannot make their own food. Instead, they make a network of threads called **hyphae** (pronounced hi-fee) which break down food. The hyphae can swap food, water, and even messages with tree roots.

Fungi help trees by cleaning up pollution and warding off other types of fungi that would make the trees sick.

Mushrooms and toadstools are the fruits of the fungus.

Douglas fir

Fungus

Nutrients

Food and water

Tree roots

In return for the fungal network, trees gift **water** and **food** to the fungi.

39

Forest family

Trees of the same species act like a family. When a seedling is struggling, its mother steps in to help. Trees that are very old, or damaged, are not forgotten. Even the strongest trees can be attacked by disease or insects and need help every now and then. By helping each other, the forest as a whole stays strong.

Tiny seedling

CARING MOTHERS

Full-grown trees take care of younger and older trees. Many young trees sprout directly below the mother tree. But the youngsters don't grow well in her shade. The mother keeps them alive by passing them sap and nutrients until they are tall enough to find the light.

Older trees often block light from reaching younger ones. Far from damaging them, it turns out to be helpful, as growing slowly when they are young helps trees to live longer.

Helping hand

This stump can't make food without leaves, but somehow it is still alive. The trees of the forest are feeding the old stump through their roots. This stump might even be the full-grown tree's mother.

Stumps can live for hundreds of years without leaves.

Tree senses

A tree doesn't have eyes, ears, fingers, or toes. Because they're so different to us, for a long time no one knew trees had senses. But now we know that they can tell what's going on around them.

They can sense the outside world...

HOT AND COLD

Trees can sense how hot or cold it is. Even tiny seeds know if the temperature has become warm enough for them to sprout and start growing.

TASTE

When an animal nibbles on a leaf, the tree can taste the animal's spit! Trees can even tell different animals apart from the taste of their spit.

TOUCH

Tree roots are amazingly sensitive. A tree can tell which roots in the underground tangle are its own. It can also tell whether its neighbours are the same type of tree.

HEARING

Roots can hear the sound of running water and grow towards it. Even when it is completely sealed off, the trees know the water is there. We don't yet know how they do this.

We are only just beginning to uncover all the mysteries of what trees can sense.

SEEING

They may not have eyes, but trees can certainly sense light and grow towards it. Every leaf on a tree can tell which direction the light is coming from. After all, they need light to make food.

TIME

Trees live life in the slow lane, but they do keep track of time. In spring, trees can sense the days getting longer. In autumn, they know the days are shorter, and prepare for winter.

...and even talk back.

BEING THEMSELVES

No two trees are exactly the same. Even if they're in the same conditions, trees will grow into different shapes. Some trees take part in the wood wide web, while others are loners.

TALKING

Trees can talk to insects using sights and smells. The colours and scents of flowers are like neon signs telling bees and butterflies that food is on offer.

SHOUTING

When they get thirsty, trees start yelling! If water can't flow from the roots to the leaves, the trunk starts to vibrate. That's a tree's way of complaining.

The **handkerchief tree** from China has flowers that look like dangling napkins.

Scientists are beginning to research how trees "think", using their roots like a brain.

43

Tree defences

Imagine if you were a tree and an insect started to nibble on you. Ouch! Luckily, trees have lots of clever ways to stop insect invaders.

Lodgepole pine beetle

Sticky sap
Beetles attacking **lodgepole pines** often find themselves in a sticky situation! Before the yummy bark can be devoured, the pine tree imprisons the insect in a very sticky **sap trap**.

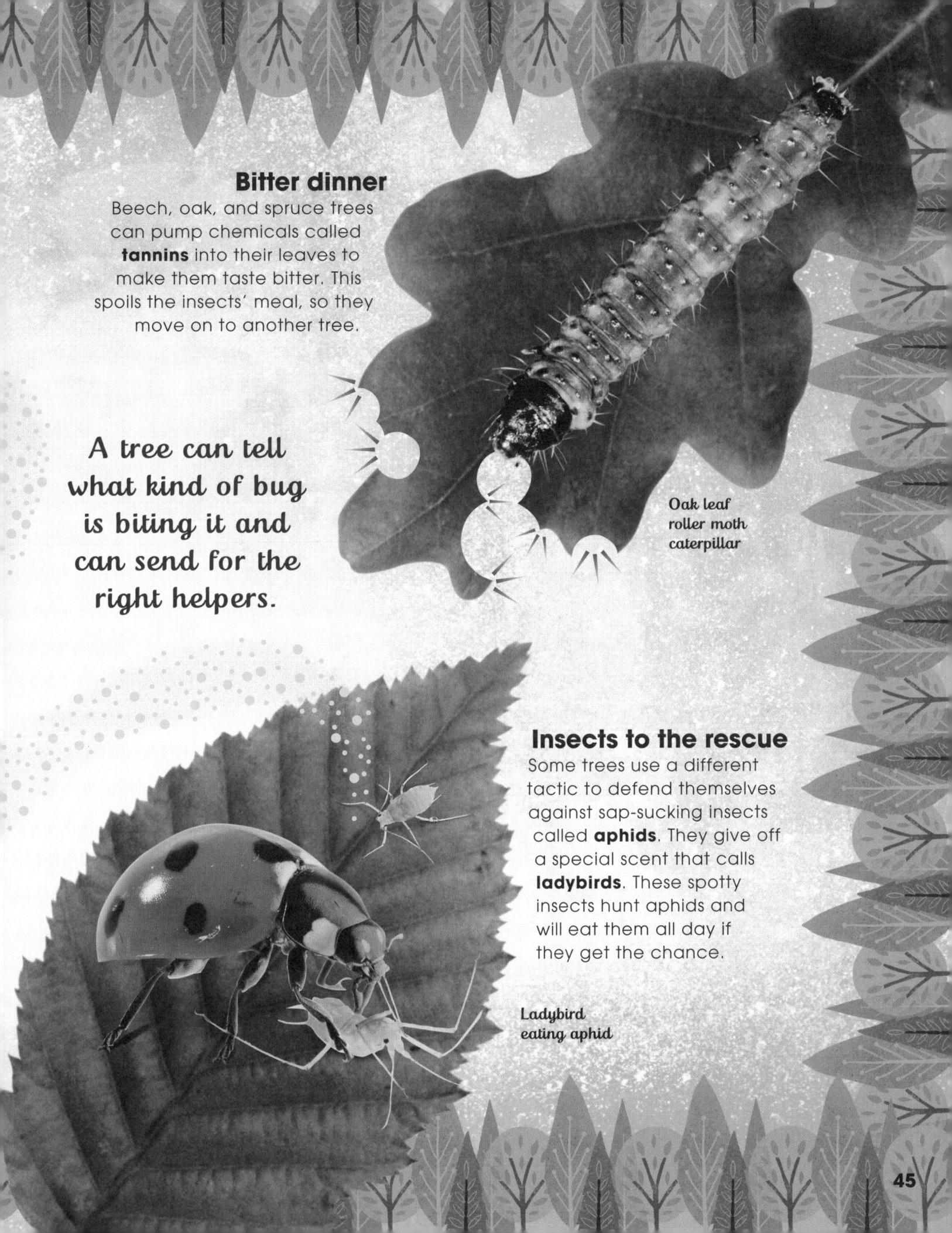

Bitter dinner

Beech, oak, and spruce trees can pump chemicals called **tannins** into their leaves to make them taste bitter. This spoils the insects' meal, so they move on to another tree.

A tree can tell what kind of bug is biting it and can send for the right helpers.

Oak leaf roller moth caterpillar

Insects to the rescue

Some trees use a different tactic to defend themselves against sap-sucking insects called **aphids**. They give off a special scent that calls **ladybirds**. These spotty insects hunt aphids and will eat them all day if they get the chance.

Ladybird eating aphid

45

Acacia thorns

Under attack

Giraffes are the natural enemies of the acacia tree. Luckily for the acacias, they have some clever defences at the ready. Not only do they protect themselves, they also **warn other trees** of the danger.

Danger!

Acacia trees can sense giraffe spit. When the tree's leaves are being nibbled, it starts making chemicals that make the leaves taste bitter. Yuck!

Beware giraffes

Acacias have long, sharp thorns to ward off animals, but the giraffe's long, bendy tongue dodges the thorns to pluck tufts of green leaves.

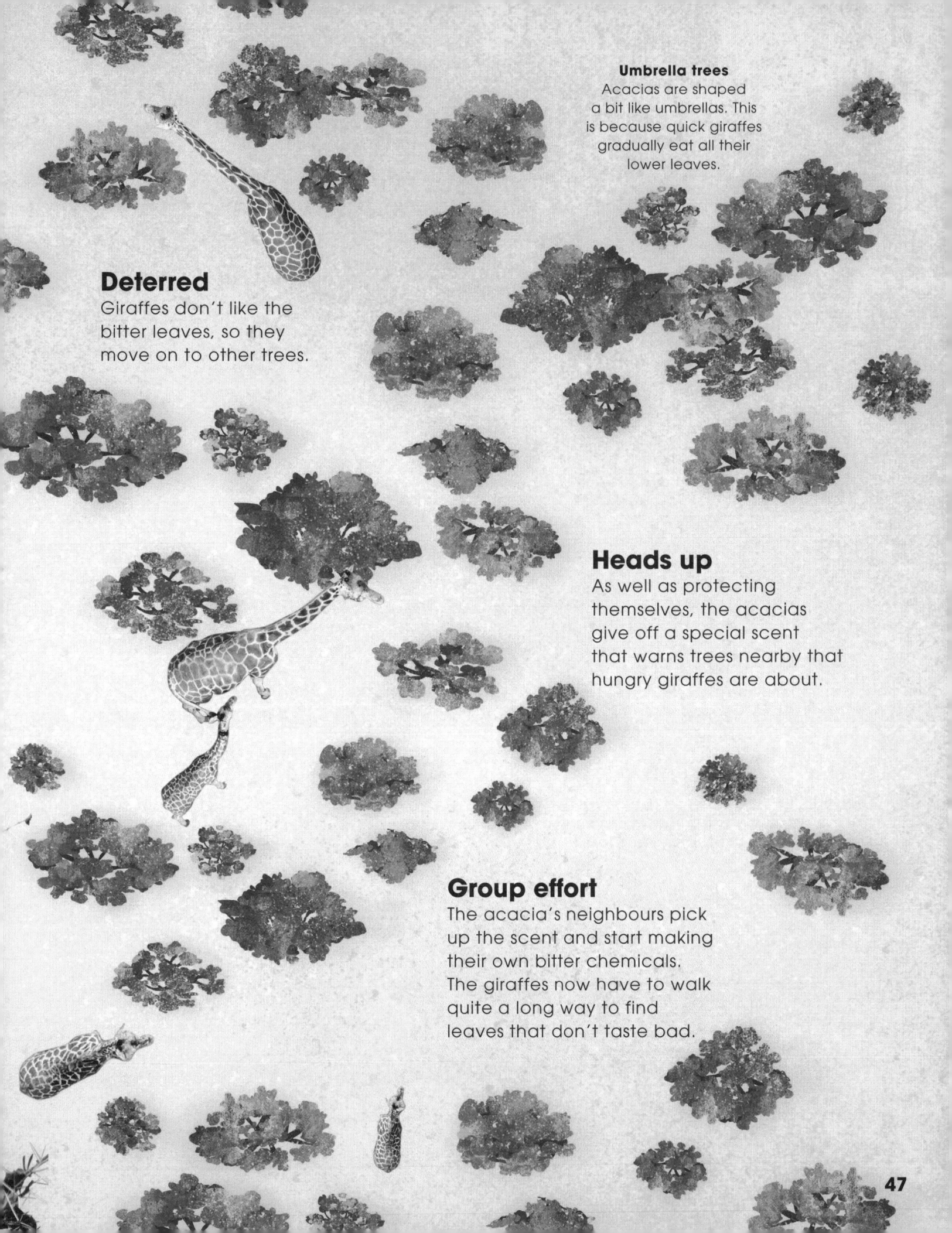

Acacias are shaped
a bit like umbrellas. This
is because quick giraffes
gradually eat all their
lower leaves.

Deterred

Giraffes don't like the
bitter leaves, so they
move on to other trees.

Heads up

As well as protecting
themselves, the acacias
give off a special scent
that warns trees nearby that
hungry giraffes are about.

Group effort

The acacia's neighbours pick
up the scent and start making
their own bitter chemicals.
The giraffes now have to walk
quite a long way to find
leaves that don't taste bad.

Incredible trees

Trees are the biggest, heaviest, and oldest living things on the planet. They can live in the snowy heights of the Andes and even clone themselves.

Tallest tree

Biggest tree

Widest tree

Oldest tree

The world's very tallest tree is a **redwood** growing in California, US. Named Hyperion, it soars 115 m (377 feet) high – the height of about 20 giraffes!

A **giant sequoia** named General Sherman is the world's largest individual living thing. It has a massive trunk 8 m (26 feet) across.

A **Montezuma bald cypress** called the Tule Tree stands in a churchyard in Mexico. It is the world's widest tree and measures 36 m (119 feet) around the trunk but is only 35 m (116 feet) tall.

The world's oldest known tree is a **bristlecone pine** named Methuselah. It grows high in the White Mountains of California, US, and is 4,800 years old.

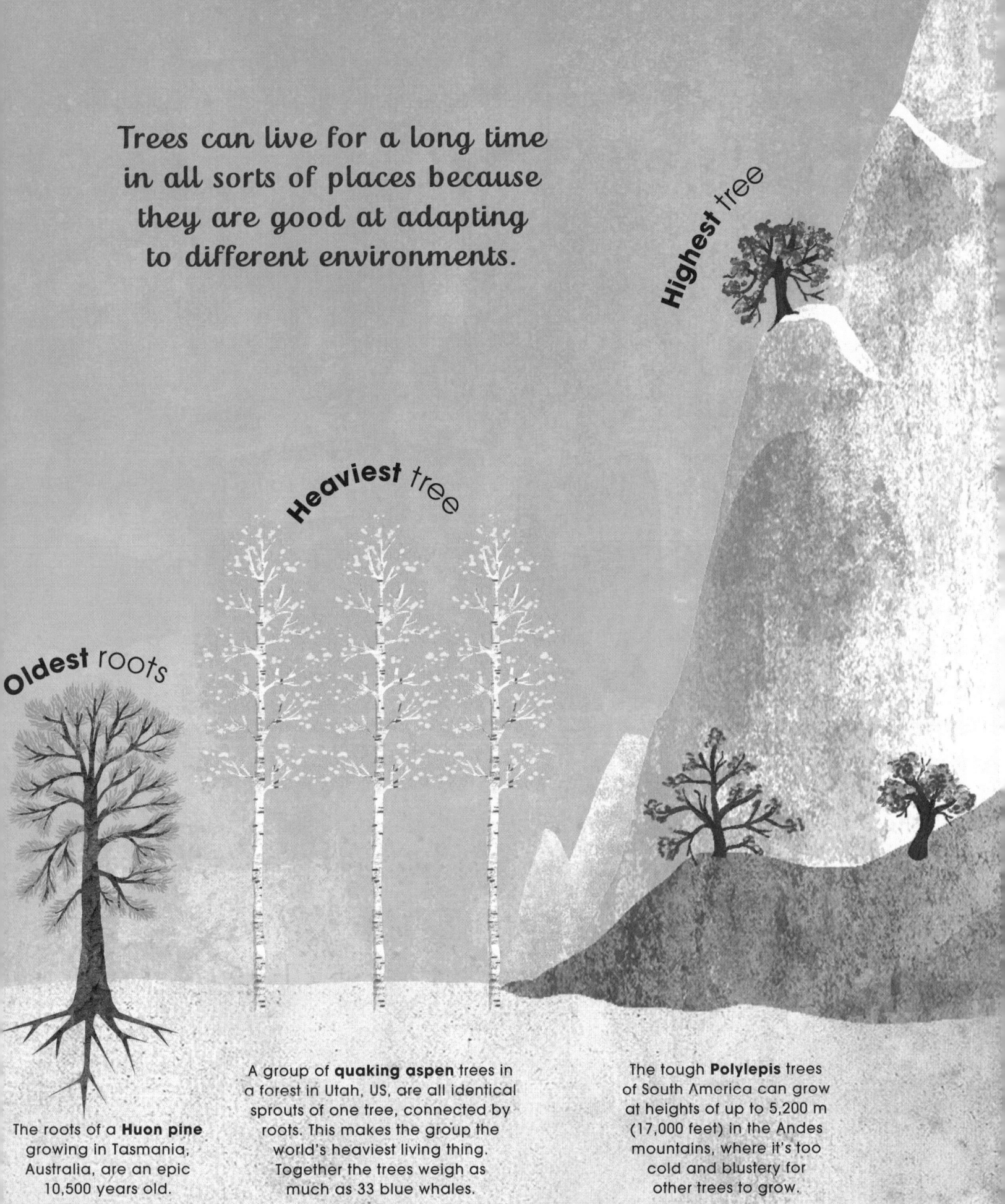

Trees can live for a long time in all sorts of places because they are good at adapting to different environments.

Highest tree

Heaviest tree

Oldest roots

The roots of a **Huon pine** growing in Tasmania, Australia, are an epic 10,500 years old.

A group of **quaking aspen** trees in a forest in Utah, US, are all identical sprouts of one tree, connected by roots. This makes the group the world's heaviest living thing. Together the trees weigh as much as 33 blue whales.

The tough **Polylepis** trees of South America can grow at heights of up to 5,200 m (17,000 feet) in the Andes mountains, where it's too cold and blustery for other trees to grow.

TREE HABITATS

Trees aren't that fussy about where they live. All they need is sunshine, water, and a little soil in which to spread their roots. That's why forests are found all over the world, and why such an amazing array of animal and plant life is found within their depths.

Trees are tough survivors. Scattered trees can take root on steep, stony mountains. They cling to life on storm-battered coasts and survive howling winds in the snow tundra. They survive on the edges of dusty deserts and on busy streets in city centres. In fact, trees can get by almost anywhere.

This section explores habitats the world over. Read on to discover how trees provide a home for an incredible variety of mammals, birds, insects, and fish.

Tree homes

This great big oak tree is home to lots of creatures. From the topmost branches to the deepest roots, animals and plants live side by side, just like people in a block of flats.

Tuck in!

An old tree like this provides food for everyone – a bit like a natural **supermarket**.

Carnivores, such as **owls and bats** are keeping a look out for small animals and insects to hunt.

Herbivores, such as insects, birds, and deer, munch on every part of the tree. They nibble on its leaves, buds, fruits, and even bark.

Below ground, in the tree's basement, **worms**, **beetles**, and **fungi** feast among the roots.

52

Blackbirds build cosy nests high in the branches to lay their eggs. The nests keep the chicks warm and safe from **predators** such as foxes.

This **woodpecker** taps the bark with its super-sharp beak to dig out juicy grubs. Then it slurps up the insects with its sticky tongue.

In autumn, **squirrels** bury nuts in the ground to save for winter.

Foxes use their sharp claws to dig snug **dens** among the tree's roots.

53

Tropical rainforest

The steamy Amazon rainforest is home to more living things than any other place in the world. A small patch of rainforest can contain hundreds of types of trees.

EMERGENT

Tall trees called emergents poke their heads above the forest canopy. They can enjoy the sun without having to compete with other trees for light.

CANOPY

Rainforest trees love the hot, steamy weather. They grow tall, creating a dense and shady roof.

Kapok trees are common in the rainforest. They can grow up to 70 m (230 feet) tall.

Green-winged macaw

Red-billed scythebill

Creeping vine

Bold beaks
Toucans have brightly-coloured beaks up to 19 cm (7.5 inches) long.

Blue morpho butterfly

UNDERSTOREY
The understorey is dim because the canopy above blocks most of the light.

At home on a tree
Epiphytes are plants that grow on other plants. Rainforest trees support many epiphytes.

Hoatzin

Two-toed sloth

Epiphyte

Long limbs
Spider monkeys swing through the trees using their long arms, legs, and tails.

Pygmy marmoset

Palm

FOREST FLOOR
The ground level in a rainforest is dark and dry because the higher levels absorb almost all the light and rain.

Giant anteater

Bromeliad

Stealthy cat
The jaguar's spots hide it among the speckled shadows of the leaves above. It slinks through the trees as it hunts.

Emerald tree boa

Tarantula

Purple harlequin toad

Golden dart frog

Pitcher

Centipede

Anaconda

Temperate rainforest

Unlike tropical forests, **temperate rainforests grow in areas of mild weather**, where it is neither too hot nor too cold. Many of the trees are conifers. This misty forest is home to lots of animals that like to keep cool.

Giant redwood

River beavers
Beavers build dams across forest streams. They gnaw through **saplings** (young trees) with their sharp front teeth. Timber! The tree crashes down to become part of the dam.

Fireweed

Big cats
Pumas are large cats that silently slink through the forest. They pounce on animals as large as deer, and kill them by biting on their necks.

Western skunk cabbage

Tree frogs
Pacific tree frogs live in forest ponds and damp ditches. They are brown, grey, or green, but can **change colour** to blend in with their surroundings.

Feel of the forest

The world's largest temperate rainforest grows along the west coast of North America. It is made up mostly of **conifers**, including the tallest trees in the world, giant redwoods. The trees grow so well because they have plenty of water. The rain is heaviest in winter but in the summer months **thick**, **damp fogs** roll in from the ocean.

Gentle giants

Moose are the largest type of deer. Males have **huge antlers**. They glide among the trees, munching ferns, grass, twigs, and tree bark.

Oregon grape

Cheeky chipmunks

Chipmunks are nimble little rodents that scamper through the forest searching for nuts, fruits, and seeds. They carry food back to their burrow in their bulging **cheek pouches**.

Raccoons

This scrappy mammal has a broad black stripe across its eyes, like a **bandit's mask**. It acts like a bandit too, raiding birds' nests to steal eggs.

Rhododendron

57

Swamp forest

In the hot and humid swamp, unusual trees called mangroves grow in slow-moving, murky water. In this coastal, tropical forest, the trees are washed twice a day by salty tides.

Proboscis monkey

Sundari mangroves

TREES ON STILTS

Mangrove trees have tall roots that lift the tree high above the water. These roots filter out most of the salt, so the tree can drink fresh water.

Male fiddler crab

The **saltwater crocodile** lurks in gloom, waiting for prey. The muddy water around the mangrove roots is the perfect hiding place for the world's largest reptile.

Falling leaves get snapped up by crabs.

Some mangroves move salt into their leaves, which then fall off.

Buffy fish owl

The Sundarbans in Bangladesh is the world's largest mangrove forest.

Black-capped kingfisher

The **Bengal tiger** prowls the swamp in search of deer. Its striped coat provides the perfect camouflage in the tall reeds at the water's edge.

Mudskippers are strange fish that can live out of water. They skitter over the mud using their strong front fins.

Many fish hide **among the roots.**

The mangrove's roots poke into the mud and slow down the water. This stops the coast from **eroding** (wearing away) over time.

59

A ghostly **great grey owl** flies silently above the forest, keeping an eye out for signs of prey on the ground below.

Spruce

Larch

The **crossbill** has a pincer-shaped beak with overlapping tips. It prises open pinecones to get at the seeds inside.

Bears use their sharp claws to climb trees and get at bees' nests. The bear breaks open the nest and slurps the honey.

Red crossbill

Bear

The male **moose** has huge, branching antlers, which it likes to scratch against trees.

Snow forest

The dense, dark, **taiga** forest is covered with snow in winter. It grows across Canada, Russia, and northern Europe, wrapping itself around the world like an enormous green scarf.

Cool conifers

The taiga is mostly made up of **conifers** such as spruce and fir. Conifers' pointy shape and waxy needle-like leaves help them to shed the heavy snow.

The **taiga** covers more of the world than any other forest.

Reindeer spend the winter in these sheltered forests, eating lichen from the branches and underneath the snow.

Reindeer

Silver birch

Pine martens dart along the snowy forest floor, sniffing out small animals to hunt.

Mountain hare

White world

Northern forests have short, cool summers and long, harsh winters. Snow often covers the ground for months on end.

61

Extreme survivors

Most trees prefer to live in mild, warm, and sunny places. However, some hardy trees can cope with incredibly **tough conditions**, such as extreme **heat** or **cold**, or long, dry periods of drought.

On the edge

Pine and cypress trees can grow on very windy cliffs. Powerful gusts break off branches, leaving only one side of the tree to grow. You can tell which way the winds blow by looking at these trees.

Dwarf willow

Beyond the northern taiga lies the frozen Arctic tundra. The shrub-like Arctic willow can cope with the bitter cold and deep snow. It hugs the ground, out of reach of whistling winds, and takes 100 years to grow just 20 cm (8 inches) tall.

Trees love drinking water but some have adapted to very dry environments. They snatch up every drop they can find.

Koalas don't get much goodness from gum leaves, so they save energy by sleeping for up to 18 hours a day.

Koala home

Tall eucalyptus, or gum, trees grow in dry areas of Australia. Koalas will eat only gum tree leaves.

Water tank

Baobab trees grow on the dry, grassy plains of Africa. In the rainy season, baobabs store water in their trunks to survive the dry season.

Fireproof

Ponderosa pine trees can survive fires. They have extra-thick protective bark that can grow up to 10 cm (4 inches) thick.

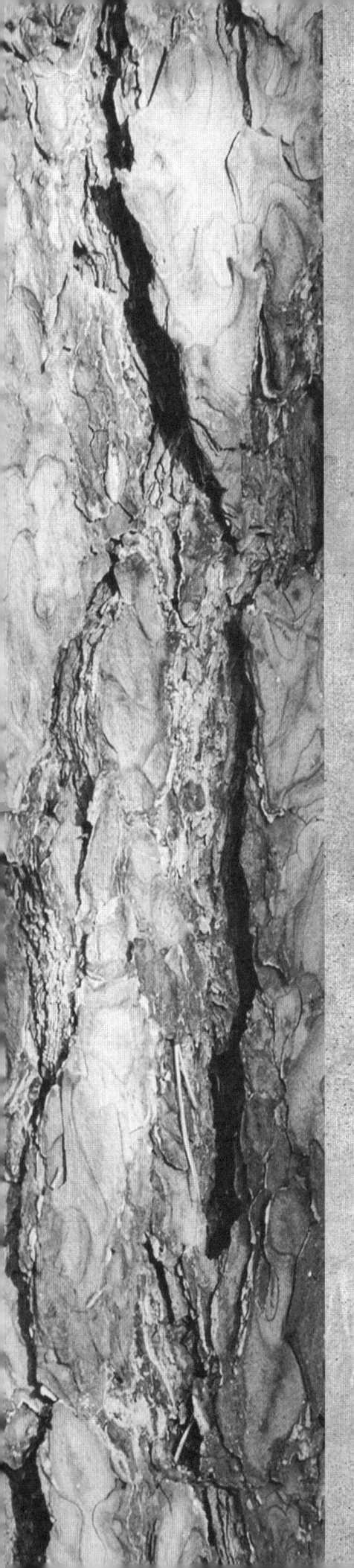

TREES AND ME

Trees help us in a hundred different ways. They provide food, wood, and many other useful products. They make the air fresh and healthy by creating oxygen, and help keep our planet a clean place to live.

Trees take good care of us and other living things. Yet all over the world, people are harming trees. We cut forests down for timber and to make room for farms and cities. We also cause pollution that is hurting trees.

Trees do a lot to keep our world green and healthy. They provide a home for countless plants, animals, and millions of people who live in forests around the world. In turn, it's up to each and every one of us to take good care of trees.

Junk food
Palm tree oil is used to make **pizza**, **biscuits**, and **ice cream**. Rainforests are being destroyed to make space for the trees that create the oil. To protect the rainforests, it's best not to eat too many treats made with palm oil.

Ice cream

Cookies

Mango

Pizza

Peaches

Avocados

Pears

Fruit salad
Fruit trees once grew wild in forests but now they are mainly grown by humans in **orchards** or **plantations**. We eat the seeds inside some fruits, such as pomegranates.

Orange

Apricots

Pomegranate

Figs

Dates

Sweet, sticky **dates** come from the date palm.

Cherries

Harvest time

From a piece of juicy fruit to a spicy curry, trees provide us with so many mouth-watering meals. Without trees, there would be no chocolate, mango yoghurt, or avocado toast! They even produce powerful medicines that help us get well when we're feeling ill.

Apples

Almonds

Aspirin

Crunchy nuts

Lots of nuts come from trees, including brazil nuts, which come from one of the tallest trees in the Amazon rainforest. Coconut palms produce the largest nuts of any tree.

Tree doctor

The headache medicine aspirin originally comes from the bark of the willow tree. Another helpful tree is the cinchona, which creates a medicine called quinine that helps treat malaria.

Cashews

Brazils

Plums

Coconut

Spicy flavours

Without trees, our food would be quite bland. Cinnamon and nutmeg can really spice up a meal! Kola nuts are used to flavour fizzy cola. Beans from the cocoa tree give chocolate its unique taste.

Cinnamon bun

Cola

Nutmeg spice on a custard tart

Chewing gum

Maple syrup

Chocolate

Sap from the chicle tree is used to make rubbery chewing gum.

Sweet syrup

The maple syrup on your pancakes is made from the sap of the sugar maple. It is collected by hammering a metal tap into the tree's trunk and catching the sap that drips out.

67

Paper goods

Paper and card are made from tiny chips of soft wood mixed with water. This makes a mushy pulp, which is then rolled flat and dried.

Wooden world

Take a look around your home. How many objects can you find that are made of wood? From pencils made from cedar to pine furniture, wood is all around the home.

It takes **at least 30 years** to grow a pine tree for wood.

Made from trees

Wood is a super-material. It's strong, beautiful, and can be grown in an eco-friendly way. We use this incredible material to build our houses and to make all sorts of objects.

The bark of cork trees is used to make pinboards, wine bottle corks, and place mats. It's also light enough to float!

Hard-wearing wood

Trees such as oak and maple make strong hardwood that is used for roof beams and furniture. Tropical hardwoods such as mahogany and rosewood are used to make guitars because they create a clear sound. Hardwood is strong and tough but the trees grow very slowly.

Rubber

Rubber trees ooze a milky sap, which is turned into car and bicycle tyres. Rubber boots, elastic bands, and party balloons are also made from rubber.

A type of incense called agarwood, which comes from aquilaria trees, is worth more than gold!

Trees and the planet

Trees are vital to every single person on the planet. They keep the Earth cool and moist, which makes it a nice place to live. Trees love to soak up water – but don't worry, they are willing to share it with the rest of us. They even help create the clouds. This process is called the water cycle.

Leaves catch rain and give off extra moisture the tree doesn't need as water vapour.

THE WATER CYCLE

Water moves between the sea, air, and land in a non-stop cycle called the water cycle. Trees play a very important part in this. By releasing rainwater back into the air, and absorbing water through their roots, they stop rain from flowing straight back into the sea.

Water rises from the ocean as a gas called water vapour.

Forests help to cool the air around the Earth. The planet is already warmer than it should be, but without trees it would heat up even more.

High in the air, the water vapour from the trees forms into clouds.

When clouds float over high areas such as mountains, water falls as rain or snow.

Rain

The clouds move inland, bringing rain to places that would be dry without them. Without trees, deserts would cover much more of the Earth.

The water flows back down towards lower ground.

River

Any rain that the trees don't absorb soaks into the soil or runs off into rivers.

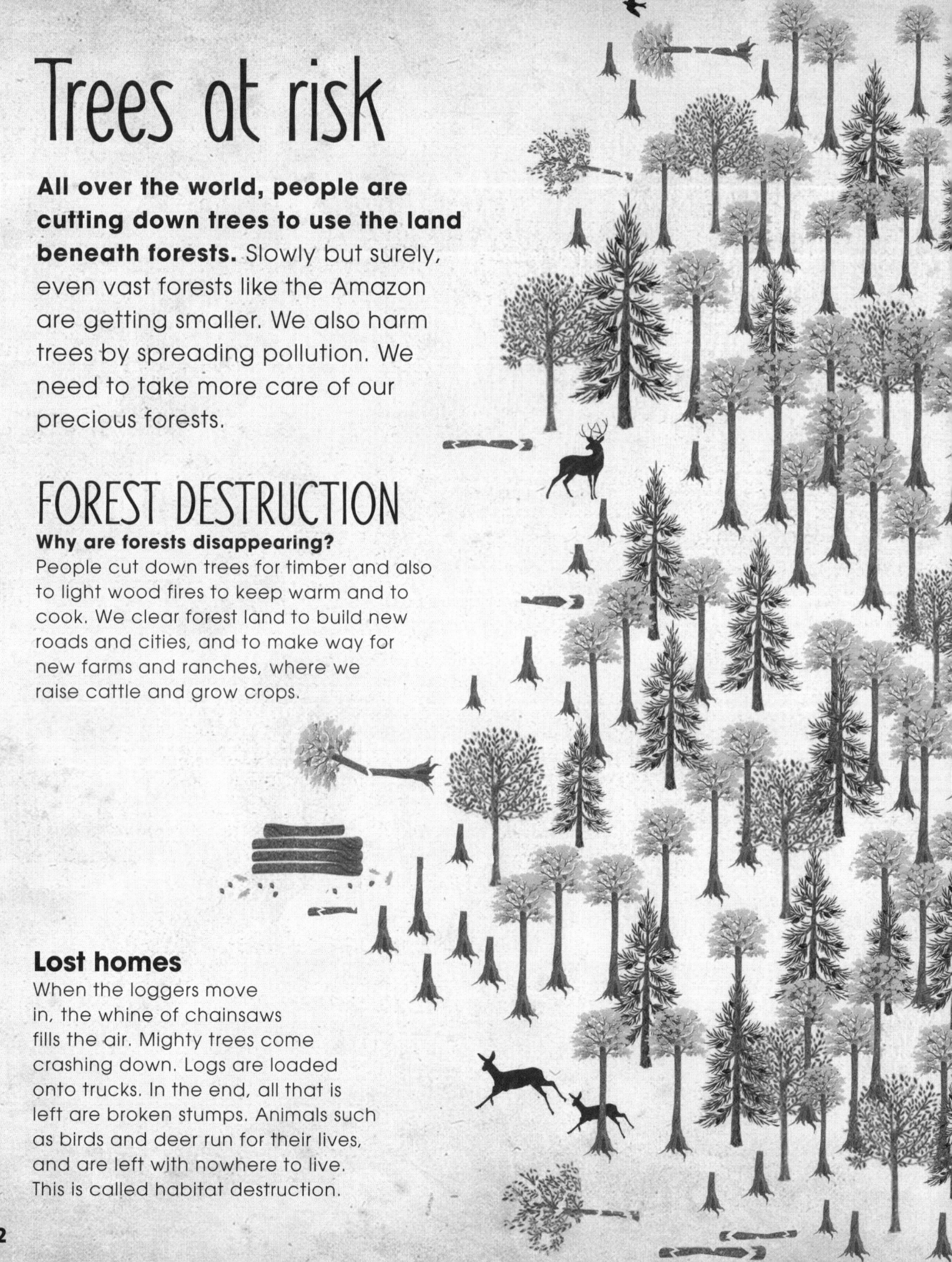

Trees at risk

All over the world, people are cutting down trees to use the land beneath forests. Slowly but surely, even vast forests like the Amazon are getting smaller. We also harm trees by spreading pollution. We need to take more care of our precious forests.

FOREST DESTRUCTION

Why are forests disappearing?
People cut down trees for timber and also to light wood fires to keep warm and to cook. We clear forest land to build new roads and cities, and to make way for new farms and ranches, where we raise cattle and grow crops.

Lost homes
When the loggers move in, the whine of chainsaws fills the air. Mighty trees come crashing down. Logs are loaded onto trucks. In the end, all that is left are broken stumps. Animals such as birds and deer run for their lives, and are left with nowhere to live. This is called habitat destruction.

Effects

Tree roots keep the soil together. Without trees to slow down and suck up the rain, it washes forest soil into rivers. This can cause flooding. Eventually, the whole area becomes drier, so farmers find it harder to grow crops.

Pollution

Cars, factories, and power plants in faraway cities can harm trees. They give off smoke and fumes that drift on the wind to pollute distant forests. The pollution makes trees drop their leaves, so they get sick and eventually die.

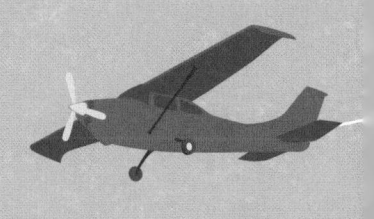

Helping trees

Trees deserve our love and care. After all, they do a lot to keep our world healthy. In turn, it's up to all of us to take better care of the world's trees. We can use recycled paper, or even plant a new tree.

PLANT A TREE

Make space in your garden, park, or other open space for a brand new tree. Buy a young tree (sapling) and make a home for it to grow.

Spade

Sapling

Dig
Dig a deep hole in the ground. Make sure the hole is twice as wide as the roots of the tree and the same depth. Different trees like different soil, but most like loose, moist ground where their roots can grow.

Plant
Plant a sapling that you have bought from a garden centre. Put a piece of wood across the soil around the top of the tree's roots. This will show you where the roots come up to. They should be just at ground level.

Some forests, such as national parks and reserves, are protected. Workers called foresters look after the trees and plant new ones when needed.

Stake

Fill
Shovel earth into the hole, making sure it is not packed in too tightly. The sapling's roots should be completely covered but with room to grow. Tie the tree to a stake to help it stand up in the wind.

Water
Water your tree to give it the best chance to grow. Remove any weeds around it and keep an eye out for pests. In two to three years, you'll be able to remove the stake and your tree will stand proud on its own.